MARTIN
LIBRARY
IN MEMORY OF
DONALD Z. HARTRANFT,
VICE PRESIDENT
BY
MOREHOUSE INSTRUMENT CO. &
HENRY A. ZUMBRUN

DEMYSTIFYING FOOD FROM FARM TO FORK

Maurice J. Hladik

iUniverse, Inc.
Bloomington

Demystifying Food from Farm to Fork

iUniverse books may be ordered through booksellers or by contacting:

iUniverse
1663 Liberty Drive
Bloomington, IN 47403
www.iuniverse.com
1-800-Authors (1-800-288-4677)

ISBN: 978-1-4620-6803-6 (sc)
ISBN: 978-1-4620-6804-3 (hc)
ISBN: 978-1-4620-6805-0 (e)

Library of Congress Control Number: 2011919552

Printed in the United States of America

iUniverse rev. date: 1/30/2012

In memory of my parents, Norman (Jiggs) Hladik and Elsceta (Toots) Hladik, who spent much of their lives on their beautiful and productive "Coulee Bend Farm".

Contents

"Farming looks mighty easy when your plow is a pencil and you're a thousand miles from the corn field."

—Dwight D. Eisenhower

The journey of food from farm to fork is complex and, when viewed from afar, may appear counterintuitive or even inappropriate to the consumer. This book is an attempt to shed some light on numerous food related activities, practices, and issues so readers might gain a broader prospective on the where, why, and how of food.

—Maurice Hladik

Preface

In North America and elsewhere, there is a growing concern by many people that they are no longer connected in any meaningful way with the production or processing of the food that they consume. The healthy reaction to this manifests itself in many interesting and useful ways, such as the hundred-mile diet, farmers' markets, urban gardening, backyard egg production, organic foods, getting to know your farmer, and other imaginative activities to enhance the connection with the food one eats. However, to make the assumption that commercial agriculture is failing us, and that the widespread adoption of some of the previously mentioned movements would be an improvement over modern farming, may not be as realistic as some proponents of these movements maintain.

Many diverse sources of information exist that portray a very negative bias regarding the production, transport, processing, and marketing of food.[1] Although there are many flaws in the food industry—as in all elements of commerce and society—there are some excellent writers who concentrate only on the negative exception and go out of their way to portray a broken farm-to-fork food industry. Some of these "armchair agriculturalists" rely on questionable science and other observations when it comes to family farms, and modern farming techniques including genetically modified crops. Common sense dictates that a book of a couple hundred pages by such individuals, without a positive word about any element of the very complex food industry that generally provides ample nourishment for billions, may be somewhat one-sided.

As someone who has had exposure to farming all his life—starting with my childhood; then as a young farmer followed by formal studies of agriculture at university; then as an international diplomat with assignments in Europe, Asia, and the South Pacific, including the position of agricultural attaché in New Zealand and Germany; and finally in an executive position with an agriculturally related business—I wish to share my insights on how cutting-edge agricultural technology and practices in the hands of today's mostly family farmers have much to offer that is positive.

In particular, I will explore the often overlooked fact that it is the responsibility of farmers and the food industry to feed a hungry world and not only cater to the informed and prosperous individuals who have the opportunity to enjoy such lifestyle choices as organic, local, or non-genetically modified foods. I will also strive to provide a balanced discussion and not back away from elements of the journey of food from the farm to the fork that could and should be improved.

Not only do farmers and the food industry meet the demands of those who wish to experience an up-market gourmet meal of carefully selected food products, but they also provide nourishment to the other end of the socioeconomic spectrum, where basic nutrition to stay alive is all that is sought. Interestingly, individuals in both circumstances may be fed from the same field.

This book is intended to entertain and inform those who wish to look at all aspects of the food industry. Some might even come to the realization that they can relax a bit and enjoy, without worry or guilt, many of the wonderful and healthy foods that are so readily available.

Acknowledgments

First I wish to acknowledge my wife, Sharon, who provided so much support and guidance on this project, including suggesting the title, *Demystifying Food from Farm to Fork.* She also patiently accepts my distracted driving when we travel through the countryside as she receives an unsolicited commentary of what I observe in pastures, fields, and farmsteads as we pass them by.

The contribution of our friends, Joanne and Wayne Shanahan, who patiently read and reread early drafts and provided valuable feedback and encouragement, was very much appreciated. I am also grateful to Teri Gentes and Louise Lettstrom-Hannant for the suggestions and friendly exchanges on a host of food-related topics that found their way into this book.

In addition, my involvement over the years with the US National Association of Wheat Growers brought me close to many of their farm family members. In particular, I wish to mention Mark Gaede and Melissa George Kessler of the Washington DC office and their contribution to my knowledge of US agriculture. There are also the countless farmers and others in the food and agriculture industry in many countries with whom I have had the honor to work with, learn from, and enjoy their company over the years.

Author's Note

I was genetically programmed to be a farmer. My ancestors for many generations—some going back five hundred years[2] in the Czech Republic, Germany, Denmark, and Norway—were all farmers. Furthermore, as an adult I have had the good fortune to personally visit all of the family's original farms in Europe except in Germany, and I still found relatives with a strong connection to these farms in both Scandinavian countries after these many generations. My two grandmothers were born in the United States on farms, one near David City, Nebraska, and the other in Norman County near the town of Ada, Minnesota. Toward the close of the nineteenth century, one grandfather emigrated from Bohemia (Czech Republic) as an infant in his mother's arms and grew up on an Iowa farm near Spillsville, where one summer during this period, Dvorak finished the *New World Symphony* and got the inspiration for *Humoresque*.

A few years later, all three families with my young grandparents independently immigrated to the same community of Wetaskiwin (which in Cree means "hills of peace") in central Alberta, where all family members over the age of sixteen were provided with a quarter section of land, or 160 acres, which in urban terms is about 16 square city blocks, or the playing surface of 160 football fields. Each family ended up with an instant and substantial landholding of quality black soil with few native trees—and thus ready for the plow and a first crop. Another grandfather came directly from a farm in Denmark as a young man, and he migrated to the same farming community where I was born two generations later.

Some of the family moved back to the United States, so I grew up visiting aunts, uncles, and cousins in Seattle, Boise, Portland, and Modesto. Although there were no relatives to visit in Nebraska, Minnesota, and Iowa, the family once took a trip back to where my grandparents had been raised. Until I was in my early twenties, my world was from the West Coast (from British Columbia to central California) to Central Alberta and Iowa, and the places in between. To me the forty-ninth parallel did not seem to exist except on maps, given the similarity of the people we knew on both sides of the border and, except for a slightly milder climate, the likeness of the Great Plains and Pacific North America, be they in the United States or Canada.

While traveling with my parents on these road trips, we usually found our way on secondary highways because we wanted to observe farmers and ranchers as they tended their fields, orchards, and herds. I can clearly recall the apple orchards and massive wheat fields of central and eastern Washington State, the intense irrigation and potato fields of Idaho, and of course the never-ending corn and grain fields further to the east. In the San Joaquin Valley of California, where my uncle and cousins had a contracting business for the construction and maintenance of the irrigation infrastructure, I had the opportunity to visit totally different types of farms where they were operating their equipment. Although I did not realize it at the time, as a very young person I had a reasonable grasp of the agriculture potential and farming activity of western North America, one of the world's great food-producing regions.

In particular, I recall my first sight of a field of corn, either in Iowa or eastern Nebraska. Where I was raised, it was much too cold for this crop, and yet my grandfather always lamented that the thing he missed most about Iowa were the bountiful corn fields. For my father and I, who also had never seen this type of farming previously, it was perhaps a farmer's equivalent of an archeologist's first visit to the pyramids.

Back at my farming community in central Alberta, the next generation of those committed to the soil involved a couple of uncles and also my father and mother, but one by one all my older cousins drifted away from agriculture, and it was a fact before I was aware of it: it was my destiny that I alone would carry on the family farming tradition. I grew up quite proud that this was my future, and because I managed reasonably well academically, it was decided that a university degree in agriculture would serve the cause well.

One thing led to another, and after six years' study and two degrees in agriculture, from one US and two Canadian universities,[3] I joined the Canadian Foreign Service, where I first served in New Zealand and the then West Germany as an agricultural attaché. Later assignments had me in Beijing, Bangkok, Hong Kong, and Seoul with a sizable staff including those who focused on agricultural and food issues. During this period I was also on loan to Agriculture Canada as director general of the Grains and Oilseeds Marketing Bureau, which also had the designation as chief advisor to the government of Canada on international grains and oilseed issues.

Upon leaving the diplomatic service, I joined the private sector[4] where for the past ten years my major activity was to scout for large quantities of biomass, particularly wheat straw in Europe, Canada, and to a very large extent in the United States, for a company with advanced technology to produce cellulosic ethanol from such material. In the United States I had the good fortune to become particularly close to the National Association of Wheat Growers, as well as the National Corn Growers Association and the US Farm Bureau. Thanks to these organizations, I met with hundreds of member farmers, often on their own property, and I developed a deep appreciation for modern agricultural practices and the approach that farmers have to advanced technology, the environment, modern business practices, their community, and of course food production, not only to feed a nearby population but to nourish the world.

As has been said before, you can take the boy out of the farm, but you can't take the farm out if the boy. All throughout my post-farm career, I have gravitated whenever possible toward work and activities that kept me in touch with farmers and agriculture.

My intent as an author was to draw upon this background and provide my views on a host of issues, practices, and activities in the food industry, starting at the farm and working through the steps to the fork, spoon, chopstick, fingers, or whatever eating utensil is used by the wide variety of people everywhere. In particular, I will place considerable emphasis on farmers' and agriculture's overall role to keep everyone on the Earth supplied with an adequate quantity of food.

1.

What Is Food?

Merriam–Webster's online dictionary food as *"material consisting essentially of protein, carbohydrate, and fat used in the body of an organism to sustain growth, repair, and vital processes and to furnish energy; also: such food together with supplementary substances (as minerals, vitamins, and condiments)."* This is a comprehensive definition that captures the technical essence of food, but I maintain that what we eat is also a personal matter for individuals, and it very much depends on the circumstances relating to the way the concept of food is entertained in the human mind.

Take a gourmand sitting down for a meal at a three-star Michelin restaurant in the company of a special friend on some summer evening in rural France. His or her attitude toward food is mostly related to a very high degree of anticipated pleasure with little thought to the above definition. This is a perfectly acceptable state of mind, as indeed the enjoyment of fine food and beverages, along with positive social interaction with others, is as old as civilization itself. In contrast, this same gourmand in quite another setting, such as finishing a vigorous workout in a hungry state, will now likely have an identity with food more along the lines of the definition. Such a hungry person, gourmand or otherwise, will in the interest of time simply make do with the offerings of a fast food outlet or whatever is handy in the fridge.

Then there are the individuals and their families who are facing food insecurity. When one is hungry and food is not available, it becomes an obsession, with all effort going into securing the next meal. Sometimes this obsession leads to acts of desperation such as theft, particularly if a caring parent has hungry or starving children.

Hunter-gatherers were almost constantly on the lookout for their next meal. For them food and its acquisition was probably the dominant activity while awake. But when the hunters brought a large animal back to the village, making a feast for all possible, then most likely the social and pleasure aspects of food became part of the occasion. While much less refined, this event would have been somewhat akin to the experience of the gourmand.

Farmers might have a different perspective. For them food may be the crop growing in the field, or many tons of grain loaded on a truck to be delivered to the market. Then there are the officials at the USDA, the United Nations, and the Food and Agriculture Organization who are concerned about the macro situation of feeding a nation or the world.

And finally, what meaning does food have for the readers of this book? It is only a guess, but many will be somewhat like me, with no food security issues but an interest in feeding the family a safe, nutritious, and reasonably priced assortment of interesting food. Socializing with meals is also important, as is a healthy interest in sustainable farming practices and the prosperity of farmers, local and otherwise. Where my food comes from and how it got to my supermarket is also of interest.

All these various approaches to the meaning of food provided a rich backdrop for me when deciding on which topics might interest a broad spectrum of readers.

2.

What Does "Farm to Fork" Mean?

As is the case with the definition of food, the term "farm to fork" can be applied in a variety of ways. For some it can involve a community picnic, where only local foods are served in order to emphasize a very direct link between food production and consumption. The emphasis here is on the local food movement.

There is also the issue of food safety as foodstuff moves from farms to the point of consumption. On IBM's website[5] they provide a description of how their technology can be applied to track food from farm to fork. The benefit of this is the rapid detection and withdrawal of any food products that have become contaminated and pose a risk to health.

Elected officials and bureaucrats use the term when drafting legislation or enforcing regulations that ensures all steps along the way are in place to facilitate the availability of ample supplies of safe, nutritious, and reasonably priced food.

For the Food and Agriculture Organization of the United Nations, the emphasis is on the socioeconomic aspects of food security and the avoidance of actual hunger. Here there is a sense of urgency and of the complexities of the often great distances from farm to fork as food is moved from areas of abundance like North America to regions of scarcity such as Sub-Sahara Africa.

And finally, when I was a boy growing up on the farm and later as a young adult, I used to speculate where the truckloads of wheat might be processed and consumed. Although I had never heard the phrase "farm to fork," I did understand the concept. The wheat could have been processed in a nearby flour mill and, against all odds, returned as bread along with our family groceries. Because most of such wheat in Canada is exported, a more likely scenario would have the grain shipped to China and the then Soviet Union, where it would be milled and converted into products quite unfamiliar to me, and then consumed in some location beyond my imagination.

To best capture all these various points of view, I turn to Dictionary.com[6] for the following encompassing definition of farm to fork:

Definition: *Pertaining to the human food chain from agricultural production to consumption*

Example: *We want to explain the complex process by which food reaches the consumer's table and the systems and technologies that ensure the quality and safety of food from farm to fork.*

I find this definition helpful because it covers almost any issue that has a direct or indirect impact on all elements of the human food chain, including the socioeconomic consequences related to consumption such as improper diets, food scarcity, or even hunger and famine.

3.

Today's Frustrated Hunter-Gatherer

Though there were neither farms nor forks during the era of the hunter-gatherers, the parallel "hand to mouth" existence would have been very real. I venture that humans were then much more connected with their food than they are today, and here lies the concern some modern humans have with the apparent lack of transparency and understanding of the farm-to-fork dynamic.

According to most anthropologists,[7] Homo sapiens have been around for between 250 and 400 thousand years. With carbon dating, they have discovered that primitive agriculture started to evolve about ten thousand years ago. If human existence on this earth were measured in terms of a twenty-four-hour clock, the earliest efforts at farming began at 11:00 PM, symbolically at the eleventh hour.

Taking the most recent date for the evolution of our species, there have been about ten thousand generations that survived by their skills and efforts as hunter-gatherers. Following this there were about four hundred generations that, until very recently in terms of human development, depended on subsistence farming that was probably augmented by a continuation of hunting and gathering. For all these generations, humans lived by their wits to glean enough food to at least keep them alive and to procreate. This quest for food was the ultimate test of survival of the fittest, swiftest, and smartest and was a constant challenge for our ancestors.

This was the most fundamental human preoccupation, and perhaps subconsciously, many individuals in advanced societies sense that something is amiss when suddenly (in terms of the long history of human existence) so little effort is required today to obtain more than ample nourishment without concern of hunger or famine from sources that have no resemblance to the environment of our early ancestors. For some folks, this brings an understandable and perhaps primordial feeling of discomfort with the entire food system. The subconscious may be telling such individuals that this is too good to be true, and taking food availability for granted does not serve basic survival instincts. Thus it is a natural human reaction to become more engaged in actual food production, or at least spend time becoming familiar with the process by which it finds its way to the family table.

Although humans started to grow their own food ten thousand years ago,[8] modern agriculture, which began with the introduction of the self-propelled steam engine or tractor in the late 1800s, is a relatively new phenomenon. Even with the introduction of mechanized agriculture, farms still remained quite small and labor intensive for many decades; it was not until after the Second World War that large-scale agriculture became a reality. Thus until about 1950 the majority of people still had some association with a farm through family, friends, or the community—there was still a connection for many as to where food came from, and food *looked* like it came from the farm. For most, this was probably enough to satisfy enough of the primordial hunter-gatherer instincts about securing food.

After 1950 this connection with farming has increasingly disappeared, with extensive urbanization occurring often in cities—where several million inhabitants had never even been close to commercial food production. Furthermore, the proportion of active farmers in the national labor force of most developed nations declined, so it was becoming less and less likely that there would be family or friendship connections with those in the urban community. In addition, many of the basic farming practices of earlier days have largely been replaced by advanced and sophisticated but less transparent plant and animal production technologies that lend themselves to large farming units to take advantage of economies of scale.

To continue with the twenty-four-hour clock, the change in the 1950s is now down to only a few seconds before midnight in terms of human history. It is therefore quite understandable that in recent decades many people in North America and other advanced economies are extremely uneasy about this loss of what had always been a fundamental element of human existence.

In the past, if the whole community was not constantly engaged in obtaining food, hunger or worse was sure to follow. Perhaps it is a bit of an exaggeration, but for some there might be a feeling of impending doom brought on by a relatively sudden and total absence of the need or opportunity to put any real effort into securing nourishment. The perpetually full horn of plenty that is a modern-day reality for many is perhaps an illusion that could disappear, leaving most individuals with little ability to fall back on their primitive agriculture instincts were the existing food industry were to fail.

To this we add the fact that much of our food today would probably be rejected by our hunter-gatherer forefathers as too strange to contemplate. Take beverages, for example. For the first several thousand generations of human existence, water was the only the liquid intake of consequence beyond infancy, and it contained no calories or nutrients except for a few trace minerals. There is now milk, coffee, tea, fruit juices, a great selection of sugar-laced drinks, and a host of alcoholic beverages. Might there be some negative, primitive reaction to drinking food instead of chewing it? Beyond beverages, there are the ubiquitous processed fast foods that we recognize as junk food because of their challenged nutritional qualities.

Not only is there a disconnect with where food comes from, but the senses of sight, taste, touch, and smell of much of what we eat probably send signals to the brain that further confuse the primitive urges of modern humans longing for a closer connection to what is on the fork. It is not surprising that there is discomfort with modern agriculture and the food supply and processing industry. Not only are the sources of much of our food today lacking transparency, but so many forms of it are unrecognizable as something our ancestors would hunt or gather.

Many people in advanced societies receive real comfort from growing their own food, shopping in farmers' markets, knowing the farmer who grows the food that is about to be consumed, and seeking out local products. It is also apparent that for some there is a negative reaction to modern commercial agriculture, given the scale of farming, the total disconnect of populations from actual food producing operations, and the difficulty to understand and appreciate advanced practices that are foreign to the farming of even a couple generations ago. The less than obvious origin of much of today's food only adds to the angst.

To further contribute to this confusing state, over their lifetime the hunter-gatherers probably faced many periods of hunger but survived these by being able to store body fat. This survival mechanism served early Homo sapiens very well, and indeed those who had a metabolism that was particularly efficient, by easily and quickly gaining body fat during periods of plenty, were the ones most likely to survive prolonged food shortages and have now passed this trait on to modern humans. Unfortunately, in today's world of nearly constant abundance for some countries, this survival gene serves many individuals poorly and manifests itself in the obesity epidemic. It is common sense[9] that those who go on crash diets trick their body into the famine-coping mode and therefore enhance the propensity to gain and store fat once the crash diet has ended. Not only does this survival mechanism work against people's well-being, but the all too frequent sedentary leisure that has replaced the strenuous effort of hunting, gathering, and primitive agriculture further contributes to the obesity issue. Though no one is suggesting returning to this earlier lifestyle, our bodies are designed to pack on the pounds, and for many an inactive lifestyle has contributed further to the dissatisfaction with today's farm to fork food dynamic.

However, in contrast to the above situation, the development of modern agriculture has many benefits, such as reducing or eliminating food security in many places on a global scale and providing nutrition at very low costs, both in actual monetary terms and when measured in units of hours of work required to put enough calories and nutrition on the table. If pure logic prevailed, this should be recognized as an incredible step forward. Furthermore, this situation has freed up humans to find creative pursuits in commerce, science, and the arts without stress regarding feeding oneself and one's clan. However, many thoughtful people are simply not wired to sit back and enjoy the perpetual bounty of our fields, orchards, fish ponds, and pastures. Fundamental hunter-gatherer instincts of survival leave many troubled with so little control or involvement in securing their food supply.

What I have described regarding being a frustrated hunter-gatherer is personal and has impacted my behavior over the years. For starters, I must garden to feel complete. When overseas in Thailand, I was blessed with a lush tropical garden with papayas, bananas, and mangoes; guinea hens largely living off what they found in the garden; and ducks in a pond. In Hong Kong I had tomatoes thriving in my south kitchen window on the thirty-sixth-floor apartment where I lived with my family. Moving on to Beijing, I again grew tomatoes on the south-facing window ledges of my apartment. When my wife, three young children, and I were evacuated because of the Tiananmen Square incident in 1989, these plants were abandoned for many days, but when I returned, although the foliage had withered from lack of water, I was left with the legacy of a few very ripe but still quite delicious tomatoes.

In all the turmoil of the time, that insignificant harvest gave me a symbolic and very welcome sense of gathering my food. Back in Canada I have my garden, which is described later and in some detail. I am not alone; countless others are embracing growing at least some of their own food, some for the first time and others as old hands who have long known the therapeutic and culinary benefits of this pursuit.

Shopping for food is something I enjoy, starting with the local farmers' market that my wife and I visit weekly, from the first asparagus in the spring until the cold of November when squash and pumpkins are the principle offerings. Here we both have an obvious and very direct connection with food production. There are also four fiercely completive grocery chains with large box stores in our neighborhood, all with ample fresh fruit and vegetable sections. When their weekly sales flyers come our way, I glance at them, and when shopping for food, usually on a Saturday morning, I make the rounds to two and sometimes three of them to pick up the loss leaders plus our regular purchases. I do the loss leader part without a shopping list or any memory prompts, and I pick the most efficient route between stores and make a beeline to the items I want at each stop.

My wife, who is somewhat amused by this weekly ritual pointed out that this takes something approaching brilliance when compared to my normal mental agility. Her only conclusion was that it is my hunter-gatherer instincts kicking in, and in my subconscious I am making the rounds of a familiar prairie and forest (I enjoy all forms of shopping for food) and using my wits to secure nourishment for the family. Indeed I study the nutrition guide on food packages in perhaps the same way as a primitive mushroom gatherer carefully studied what he had picked to determine if it was safe to eat.

Perhaps this book will assist today's version of a hunter-gatherer to better come to grips with the reality of the modern food industry and if there is angst regarding where the next meal really comes from, my attempts at demystifying the stages from farm to fork may prove to be helpful, interesting, and perhaps even entertaining. It is impossible to turn back the clock for entire populations and go back to the simpler life of a hunter-gatherer or a primitive farmer. I would hope that for some readers this book will provide new and reassuring insights into modern agriculture.

4.

The Family Farm—Where Food Begins

Folks who have never spent time on a modern family farm may have a legacy vision of food production, perhaps as depicted in Norman Rockwell's 1948 painting *The County Agent*. There is general uneasiness that an extremly pleasant, colorful, and imporant part of our heritage has been squandered for no good reason—or worse, because of corporate greed.

This bucolic painting depicts what is clearly a wonderful family farm lifestyle that to this today may be considered by many as the ideal source of our food. The family seems closely bonded, and by the dress of the boy, he appears to be contributing to the farming effort and probably is encouraged to follow in his father's footsteps and take over the farm some day. The chickens are not only free range but loved as well, and the calf also seems to be in a good place and is totally comfortable in the presence of humans.

Although the scene oozes a quality lifestyle, there is little evidence that the farm produces a lot of food. The calf and chickens are obviously pets, but their numbers, at least in the painting, are insignificant, and there is little evidence of any other type of farming activity. The point that I am making here is that to lament the loss of the small family farm—and to have major concerns about the so-called factory farms being responsible for the demise of such a lifestyle—is making an issue where one probably does not exist. The reality is that there is no other option but to rely on large commercial farms, which are overwhelmingly of the type that are family owned and operated, as the production unit responsible for most of the food that ends up on the fork. .

To illustrate the contrast between a bygone era and modern farming, the following photo depicts the new reality in food production. For starters, according to the 2007 United States census data[10], nearly all farms today, no matter what the size, are family owned and operated. Thus it is probable that one of the operators running the combine holds the deed to the property and likely lives nearby in his farmhouse. The other operator may be an employee but is most likely a family member.

Two Combines Harvesting a Wheat Field

Each combine can likely harvest about fifty tons of wheat per hour (the largest such machines on the market will manage seventy-five tons per hour under ideal conditions), which if converted to whole wheat bread represents close to one hundred fifty thousand one-pound loaves. Given that a loaf of bread has a little over a thousand calories, in caloric terms and for illustrative purposes, this farmer harvests enough wheat in an hour with two combines to feed about one hundred fifty thousand people for a day. In an eight-hour day, enough wheat is harvested by just to operators and accompanying trucks to provide the calories and a lot of nutrients for more than thirty-four hundred people for a year.[11]

Yes, smaller planting and harvesting equipment could be used, with several farmers operating the same piece of land that is divided up into lesser holdings, but they are probably using the same basic farming methods. In any other industry, such efficiencies in labor productivity, as the picture above depicts, would be heralded as a success. However, because of general concern that the small family farm and an ideal way of life has been cast aside to make room for such operations, this situation is considered by many as negative, rather than a blessing to feed a hungry world with high-quality and low-cost food.

In reality, for the consumer what difference does it make if the wheat content in a loaf of bread comes from a large and well-managed family farm or from a much smaller acreage, as might be associated with the Rockwell painting? Both would be nutritionally identical and probably produced with the same approach to growing wheat. The only difference is that the large farm can probably bring a bushel of wheat to market using less energy, disturb the soil less because of the use of costly and advanced planting machinery and require a lot less labor by utilizing advanced, energy-efficient, and large-scale equipment that can only be profitably utilized on larger holdings. However, the wheat would be exactly the same to the consumer from both operations.

One major development in recent years has been the advent of conservation tillage, whereby at least 30 percent of the previous crop's residue remains on top of the soil. The most commonly practiced conservation tillage method is "no till," whereby the next year's crop is planted directly over the previous harvest without significantly disturbing the soil. Unfortunately, for smaller farms the seeding equipment required for no-till tends to be capital intensive and only provides a good return on the investment if larger acreages are involved. More details on this practice will be provided in the chapter on food production technology, but the main advantages of this technology are a substantial reduction in fuel, rain and snow retention on the field, less evaporation, and dramatically reduced wind and water erosion.

Without doubt, the use of fertilizers, herbicides, and pesticides result in environmental degradation, particularly with such material finding its way into lakes, streams, and rivers. However, it should not be overlooked that these inputs are very expensive, and thus farmers are parsimonious in their use as unnecessarily applications cut into the family income. Even here, advanced global positioning systems with precision seeding and chemical application make a substantial difference. Furthermore, many farmers are as environmentally conscious as anyone, particularly because they must live in any disruption to nature that they create. Here is where larger operations can afford costly GPS-monitored equipment that reads the soil and weed conditions from satellite data and applies fertilizer and herbicides in precise amounts as required.

Perhaps no other major industry has made so many sustainability breakthroughs as farming has over the past three decades. More details on this good news story will be covered in a subsequent chapter.

My own life experience regarding the advantages of larger scale operations is perhaps relevant. I was raised on a farm where my grandparents homesteaded, and as I was growing up, we had a few dairy cows, several hundred sheep, three chicken barns with perhaps six hundred laying hens, occasionally some pigs, a few bee hives, and around four hundred acres of high-quality land with a variety of wheat and other grain crops, pasture, and forage. In most years while my father usually made some money from each operation, there were no economies of scale, and the human effort required for each unit of output was very high. When crops failed, he had to temporarily leave the farm and work in construction.

He realized that such diverse operations were less than optimal as far as profits were concerned, given the limited labor available, and when combined with my newfound knowledge from my agricultural training, the livestock operations became streamlined into beef cattle only, with grain produced for their feed. Not only did income increase by a substantial amount with less human effort involved, but a lot more food came from the same farm. This was a win-win situation for the farmer and consumer, though mixed farming with many types of food products on our farm was no more. Indeed, this has been the case in recent decades on most commercial farming operations throughout North America.

With my studies, my off-farm professional career options widened significantly, and I also became very conscious of the fact that while the family farm could provide my parents and remaining siblings with a decent income, the enterprise was not large enough to also support my wife and two sons, who had become part of my life while I was away at university.

The situation was particularly clear to me because I had written my master's thesis on the discrepancy of average farm incomes versus society as a whole. According to data that I used at the time, during the two decades after World War II, the average farmer's income was only half of his or her urban counterpart. My conclusion was that either farm gate prices had to rise substantially, or farm consolidation must take place with each farmer producing substantially more food to achieve similar incomes to their urban counterparts. Given market realities, the second scenario was the only viable one and thus I concluded that to avoid widespread rural poverty, the farm consolidation must occur with commercial operations becoming much larger.

To me, if I was to become a fully engaged farmer, the only option was to take on substantial debt to essentially double the scale of the family farm operation. This was the late 1960s, and even then a farm approaching a square mile (in city dimensions, about sixty square blocks) and a cattle operation of around three hundred beef animals was only of a sufficient scale to provide a decent income for one family. By decent income, I expected to have an average-sized, conventionally equipped modern home, a reasonably late model car, funds for the occasional holiday and other recreation, an ability to provide higher education for my family, some savings, and other basic comforts of life. I was seeking nothing more than the expected quality of life of a hard-working urban family.

I worked through the economics of achieving this middle-class lifestyle along with the financial burden of carrying a lot of debt that would be required to expand farming operations. I realized that the potential profits simply were not there to achieve the standard of living that I had set for my family. With incredible reluctance I made the decision not to return to the farm.

My father farmed for a few more years, and when time came to sell the operation, it proved too small to attract another farmer with sufficient funds who was content with such a scale of operation. He then sold the land to two local farmers with adjacent land who, unlike the other potential buyers, were quite willing to pay top dollar because each already had a few thousand acres that they had accumulated over the years. They had the machinery and other infrastructure to easily integrate a few hundred additional acres into their operation. Thus the land was a greater economic benefit to them than to someone who had to not only purchase the land but also accumulate the necessary equipment and other inputs that that were already in place for the larger farmers. One of the purchasers was the third-generation farmer whose father and grandfather were neighbors of my family for decades. Both these farms are still operated by the owners who live on the same farmsteads that have been the operation headquarters of the nearby land for nearly a hundred years.

My parent's farmstead, which included a pleasant house in a treed setting plus a few uncultivated pasture acres, was sold to a hobby farmer who, along with his wife, had off-farm jobs. I mention this because the population dynamics of the community was not affected by the land consolidation; all existing farms remained active and well maintained without any loss to the neighborhood dynamic.

As for my father, he then bought a small farm across the road that had one field and a forested stream running through it. He knew the property well and was of the opinion that this particular piece of land should never have been cultivated, and as his legacy he let native grasses, bushes, and trees establish themselves as nature intended. When he passed away some twenty-five years later, I was able to spread his ashes on a twenty-acre plot of ground that was well on its way to becoming natural habitat similar to that which existed before European settlers arrived. I also took the liberty to spread some ashes on the original family farm. It was harvest time, and a bountiful crop was in the field. The land was obviously well maintained, with a lot of wholesome wheat about to be gleaned.

There was no interference by the mythical bogeyman of agribusiness, but rather the orderly consolidation of land into family farm units owned by existing neighbors. These tend to be very efficient and well managed, and they provide a good income for the owners. Yes, these are multi-million-dollar operations, and perhaps that causes an understandable discomfort to some who opine that many would-be farmers have been deprived of a wonderful way of life by the dominance of large-scale agriculture.

This is by no means a unique story; it has been repeated untold times in the United States, Canada, Australia, and other countries with a strong agricultural base. Most in the farming community and those involved in agriculture research and business look upon this as a positive development. The land tends to be left in the hands of the most competent and financially stable farmers who have the resources and know how to best utilize the soil.

The widespread consolidation of land into larger family operations has been beneficial on two very obvious fronts. Food production is constantly setting yield records, and prices to consumers as a proportion of income is in a steady decline in recent decades. The consumers who benefit from this dynamic not only include the generally well-to-do North Americans but also many who might otherwise face food security issues, both in our own society and in distant and less prosperous countries.

Apples, which are mostly produced on family run orchards, provide a good example to demonstrate just how much food is delivered to larger urban centers. Three apples weigh about a pound, which translates into about six thousand apples in a ton, so a loaded large semi trailer holds about a hundred fifty thousand apples. To supply a mid-sized city, such as the greater Denver metropolitan area of 2.5 million, with one apple per person, this would require over sixteen large truckloads of the fruit.

To provide context on the massive flow of food through the system, the average consumption of apples and apple products in the United States is about 50 pounds per person per year, or 150 individual pieces of fruit. This amounts to a thousand truckloads per day to provide the nation with an average of half an apple per person, or about 365,000 truck shipments per year.

These are most definitely mind-numbing numbers, but they are provided to demonstrate the immense scale of operation to provide one food item to over three hundred million people. Large-scale production, efficient storage, distribution, and marketing operations are the only practical means to make this happen with a consistent supply year after year at reasonably stable prices.

It could be argued that many times the number of small apple growers could also achieve this level of production, but what would this do for costs with each one supposedly deserving a decent middle-class standard of living for their families? Also, there would be the challenge of managing the relationships with several times the number of farmers and coordinating their deliveries to processors and packers. The more players there are, the more costs that arise, and that cost must be passed on to the reluctant consumer, who gains no benefit from apples originating on smaller orchards.

Quality control issues would also arise. Though consumers want variety in the food they can purchase, they do not like variation. For apples, if it is a red delicious, all must be about the same size, shape, and color. Blemishes are shunned by consumers and often lead to premature deterioration. Larger growers with large, well-managed orchards and the latest in expensive apple-handling equipment can meet this challenge with much greater ease than many small and diverse apple farmers.

For the prosperous and well-connected consumer, such a massive industry may be of negative appeal because they have the wherewithal to source whatever food they want at any time of the year. However, the existing approach to cost-efficient, large-scale farming and food distribution is essential to provide value-priced and quality products to millions of people who have neither the financial means to pay up-market prices nor the time and transportation flexibility to seek out other alternatives for their nutritional needs.

Those with financial and other resources at their disposal, and who are concerned with this scale of the food industry, have the option to circumvent such sources through farmers' markets, adhering to the hundred-mile diet or other similar food procurement strategies. These sophisticated food acquisition tactics may be an admirable approach to putting food on the fork, but it is not helpful to be critical of modern commercial agriculture and the benefits that it provides to so many who are only concerned about achieving the best possible return for their food dollar.

Moving on to another farm-related topic, a perceived major negative factor with such large –scale, food-producing operations is the impact of the population decline on smaller rural communities. This in some instances has certainly been disruptive to the way of life to those left behind, with schools being closed and consolidated and the inconvenience of access to medical attention and basic shopping being challenged as well. As an example on how farm consolidation can also play out with a more benevolent end result, I like to use the example of Imperial, Nebraska,[13] in Chase County, where I had the opportunity to visit on three occasions as part of my previous employer's quest to locate biomass for conversion into fuel ethanol.

Blessed with excellent farmland and water for irrigation, Imperial is located in the southeast part of the state near the Kansas and Colorado borders, and it is about as rural a community as one could find based on the distance from a major city, 200 miles from Denver, 350 miles from Omaha, and 20 miles from Interstate 70. However, with a population of around two thousand and nearly the same within a fifteen-mile radius, the town has its own superb dynamic. For example, when the local movie theater no longer became a profitable business venture, instead of having it close permanently, a group of enlightened citizens took over the operation on a volunteer basis and maintained the practice of showing late-release movies along with the usual popcorn (incidentally a major local crop) and cold drinks a few nights a week. There is also is a weekly newspaper, a country club, two golf courses, a swimming pool, over five hundred students in K-12 with a ten-to-one student teacher ratio and above state average ACT scores, a community college, eight eating establishments, a little theatre group, a county fair, active sports teams that participate at the state level, a municipal airport, and a host of companies that cater to the requirements of the local but large-scale farms.

The population density of surrounding Chase County[14] is only four per square mile (Manhattan is nearly 70,000 per square mile). The large, mostly third- and fourth-generation family farms average 1,655 acres (over 1,600 football fields) with sales averaging $436,000 per year. After expenses the national average for farm income was slightly under $40,000—not very much considering that the value of the average farm, including equipment and machinery, would exceed a million dollars. The main crops in Imperial are wheat and corn, which account for nearly half of the gross farm income, while livestock make up most of the balance.

I have provided considerable details for one small but dynamic community to illustrate that rural life after farm consolidation remains vibrant even if the population density is very low. Furthermore, although the farms are large with substantial cash sales, the profit level would not indicate that farming in this community, which is typical of hundreds more throughout North America, comes close to the stereotyped corporate farms that are accepted as reality by so many current writers who condemn the current food production system.

Continuing with rural America, let us go to southeast Idaho, the center of US potato production, which remains mainly in the hands of local farmers who have been farming in this corner of the state for generations. I visited this region many times because there was abundance of wheat and barley straw (biomass for my employer), given that this was an essential rotation crop along with nitrogen-fixing alfalfa for potatoes. I raise this agronomic fact to illustrate a sustainable and widespread practice of crop rotation, a fact that is often overlooked by those who are critical of modern agriculture.

A couple incidents during my visits come to mind that illustrate the typical sophistication, character, and humor of the family farmers. On one occasion I was bouncing along a country road in a cluttered and very dusty car with a seed potato grower, when he stopped the vehicle and fished out a laptop computer from under the driver's seat. He advised that it was time to check his half-dozen storage buildings, each of which held several hundred tons of seed potatoes. He explained that potatoes stored best in a very narrow temperature range over winter, slightly above freezing. The buildings were controlled by fans moving air through the storage area. I was quite taken aback because he electronically moved from building to building (which were located about fifty miles away) to either confirm that things were as they should be or tweak settings a bit to hit the right temperature. When I expressed my amazement on how farmers such as he had adopted advanced technology, his reply was, "Just call me a high-tech redneck!" Incidentally this rather large operation was also farmed by his sons, brother, and nephews.

On another occasion I was visiting a farm located several miles north of Idaho Falls. The owner showed a Royal Dutch Shell executive from the Netherlands and me around to demonstrate that this was a promising community with ample biomass for cellulosic ethanol production. The conversation about his farm and the farms in the area turned to the problem he was having with rattlesnakes and his calf operation. He went on to explain that during this time of year, the rattlers would be in their dens in the ground, and the best way to eradicate them was to place dynamite far down into their burrows, and they would either die from the blast or never find their way out. An invitation was then extended for all of us of us to give it a go that afternoon. While I was ready for the adventure, the Shell executive made it very clear that he would have nothing to do with such an activity. His reluctance probably had something to do with the fact that Shell, a very safety-conscious organization, just might include both dynamite and rattlesnakes as things to avoid, particularly when combined. As it transpired, our farmer friend was pulling our leg just to get a reaction and to bring levity to the conversation.

Then near Stuttgart, the rice capital of Arkansas, I had the opportunity to join a local farmer in the air-conditioned cab of his state-of-the-art combine one fall day to harvest part of his three thousand acres of top quality long grain rice. That day, as was the usual routine for that time of the year, he was working with his two brothers. One operated another late model and very large combine that, like the one I was on, had a value of around four hundred thousand dollars. There was a healthy competition going on between the two operators on who could harvest the most rice per hour in an efficient manner that left a minimum of grain behind. The third brother was responsible for coordinating the trucks to haul the rice from the field to the mills in Stuttgart. The father, who no longer operated heavy farm machinery but was still very active in the farm's management, visited us in the field to determine how harvest was proceeding. This farm, while very large, was typical of so many family farms that I visited over the years as part of my previous job.

On another occasion I asked a North Dakota grain farmer how long his harvest took, and his response was that it took twenty-nine movies. With advanced GPS navigational equipment, combine operators monitor the machine much as an airline pilot does on autopilot. He thus had the luxury of enjoying movies in an air-conditioned combine cab while gleaning enough food for many thousands of people, not only in the United States but throughout the world. This is very sophisticated, expensive, and high-capacity equipment—but operated by a farmer and not a manager of a so-called factory farm.

There is concern in some quarters that farmers even if they are owner operators, they are so heavily influenced by outside corporations that they have lost all flexibility in how they farm. While they have the same contractual arrangements with bankers, suppliers and others as does any operating business, the decisions on what food to produce and where to sell it are generally are those of the farmer alone. Also if they are not pleased with the service of bankers, machinery dealers, and the farm supply companies that provide fertilizers, chemicals and other inputs, there is ample competition for them to easily seek alternatives that offer better process, products or service. While some farm products are produced under contract to a particular company, these are normally annual agreements with the farmer free to either change what he produces or switch to an alternative marketing arrangement.

Returning to my North American rural travels, in Canada I visited the community of Birch Hills,[15] which is a rich, grain-growing area (and therefore biomass producing area) in Saskatchewan just south of the Canadian boreal forest borders. It has about half the population of Imperial but is similar in makeup as a vibrant community surrounded by equally large family farms. There are four eating establishments, all local and no franchises, with one that is quite outstanding. Jenni's New Ground Café[16] features a new menu each day from local products whenever possible, live music some evenings, and walls that serve as a gallery featuring the work of local artists.

Birch Hills also regularly hosts the provincial crokinole championship, a board game somewhat unique to Canada that takes considerable finger dexterity to place the wooden shot pieces in a winning position. Some readers may think this as a rather quaint pastime, but it required immensely more skill and comes with a lot more sociability than the players of the slot machines might enjoy.

Indeed it is not so much the game but the gathering of like minded individuals in a vibrant farming community that is at play. I recall my own childhood, when there were weekly neighborhood whist (simplified bridge) parties during the winter that rotated from house to house. Then, as well as now, the community spirit resonates in rural North America and wherever there are farmers.

In addition to time spent in a host of communities, I have also been a regular presenter on biofuels and biomass production at major conferences organized by the National Association of Wheat Growers and the National Corn Growers Association. Thousands of farmers from all over the nation attend these occasions. For example, in 2008 Commodity Classic, the joint annual farm show put on by the associations representing the US wheat, corn, and soybean growers was held at the Gaylord Opryland Resort in Nashville, one of the largest hotel complexes anywhere. Growers who filled the place were sophisticated and generally prosperous folks, but they were still farmers who lived with their families on their land. This event was attended mainly by couples plus their children, who were being groomed as the next generation to take over the family farm.

The above vignettes of the farmers and the communities hopefully provide readers with an appreciation that rural North America is a vibrant place, not at all dominated by corporate farming. Here most decisions about primary food production are made at the kitchen table by the same people who work the land and not in a boardroom in some remote urban location.

To back up my observations of the family farm as the core unit responsible for most food production in North America, I turn to the USDA statistics drawn from the most recent census. In USDA/Census Bureau[17] terminology, "Individuals/family, sole proprietorship (farms)" accounted for an identical 86 percent of all farms in the United States in both the 1997 and 2007 census. At the other end of the spectrum, only 0.5 percent—or one farm in two hundred—is classified as a "non-family corporation (farm)."

The communities and individuals described above do not match the generally held contemporary perception of modern commercial farming with a landscape dotted primarily by so-called factory farms managed by profit-driven, industrial-type managers, who do not pay sufficient attention to sustainable farming methods and acceptable animal husbandry practices.

Beyond communities such as Imperial, which has adapted well when extensive farm consolidation reduced the local population, what about the entire rural landscape? Somewhat surprisingly, when comparing census years on a national basis, the total rural population including farmers and others outside of urban areas has grown by 10 percent from forty-five million in 1980 to over fifty million in 2008. This is approaching the population of France or the United Kingdom and represents one in six Americans. Also, the number of active farm operators has been at a near constant 2.2 million between 1997 and 2007. Interestingly—and probably not widely appreciated—women are a growing proportion of farm managers and have increased from around 10 percent to 15 percent over the ten years.

Small farms still dot the landscape with 54 percent of all farms, or well over a million holdings in the one to ninety-nine–acre category, according to the 2007 US census. Indeed the decade has seen a fairly substantial increase, from 49 percent of the total in 1997. Small-scale farms are far from disappearing.

These, the most public and reliable of population statistics, seem to be describing another planet and are totally at odds with the picture of rural America that is depicted by so many writers and speakers that lament the demise of the family farms in favor of large-scale corporate agriculture. The demise of rural America, and indeed the family farm in the nation, seems to be somewhat of an urban legend.

There is a frequently overlooked downside that the operators of these small family farms face, in terms of a decent livelihood from farming. Again using the 2007 census, nearly 60 percent of all farmers report sales (not net income) of less than ten thousand dollars per year. This represents an increase in the number of these low-income food producers from 1997, when 55 percent of all farmers were in the category. Sales, of course, do not mean income, because there is a varying degree of out-of-pocket expenses to produce the food that is sold into the market. In other words, over half of American farmers have a net profit of well under ten thousand dollars per year.

For many, farming is a hobby that pays a little, and these folks are probably in a good place with other income from a well-paying day job, perhaps in a profession. For others, after a full day's work, they put in extra time to earn an additional few dollars from the farm. This is a less desirable set of circumstances, but if such part-time farmers enjoy this lifestyle, both the individual and consumers benefit.

Finally, there are the rural poor who eke out a life on the farm and depend on the few thousand dollars a year for much of their livelihood. This type of family farming operation represents a social issue affecting those with minimal skills for off farm work, too little capital to provide a critical mass of food production, and perhaps insufficient experience in agricultural techniques to make the most of what resources they do have. Similar unfortunate circumstances can be found with individuals engaged in most types of commercial activities in other parts of the economy and should not be seen as a result of serious flaws in the food production system even if the activity fits the stereotype of a bygone farming era.

The 2007 census the USDA points out that of all farmers 65 percent work off the farm and only 45 percent indicate that farming is their primary occupation. For new farmers who started farming between 2003 and 2007,

79 percent worked off the farm and 33 percent were full time farmers.[18] Thus a majority of those who operate a farm have a second or perhaps even a third source of non agricultural income if there is an employed spouse

Many of the organic produce growers and small farmers, who garner so much support from their urban counterparts as the perceived desired backbone of American agriculture, are either hobby farmers, overworked folks trying to make an extra buck after a nine-to-five job, or the rural poor. All three groups should be commended for their tenacity and hard work on the land, but they do not constitute a viable alternative to the larger family farms that provide the bulk of our food.

Furthermore, while would-be farmers are free to enter the world of agriculture on a small scale, the encouragement to do so should be tempered with the realization that the rewards are mostly the pleasure that producing food provide but the income earned, and the impact on the total food supply will be modest. If the very real lifestyle issue is sufficient to encourage folks to become farmers, then by all means they should take up the challenge. Indeed, as has been mentioned previously, the farmers near Imperial only made a profit of forty thousand dollars in spite of nearly a half million dollars in sales. A farming operation needs to be very substantial to return an income comparable what is commonplace in an urban setting.

To capture the urban perception of modern agriculture, I borrow and adapt Mark Twain's response after learning that his obituary had been published: the reports that the family farm is dead are greatly exaggerated. What is exaggerated it the ideal that so many have regarding the small family farm and the belief that our nutritional needs would be much better served if there were many more such operations.

Moving back to larger farming units, there is a unique operational dynamic in favor of family farms when compared to the normally successful conventional corporate management approach to running typical business. Most types of food production are influenced by the vagaries of weather, intense but frequently disrupted planting and harvesting operations, and changing market conditions; all of which the farmer has little or no influence, control or knowledge of what lies ahead. Yet decisions must be made, and farming action is taken with such imperfect knowledge, often several months before harvest and the ultimate sale of the product of their efforts and risk exposure. This risk-prone scenario is something farmers have come to live with but is totally outside the normal comfort zone of the corporate world. This is probably the principle reason why only one farm in two hundred in the United States is classified as corporate.

A corporate management structure reporting to a board of directors representing the interests of outside investors simply find the risks and unknowns of many types of farming too great in the context of modern business practices. However, certain types of livestock and poultry operations, where activities are unaffected by seasonal and weather fluctuation, lend themselves to more of a factory type production routine with more predicable outcomes, and the therefore work quite well under corporate-type ownership and professional management.

In addition to the world of risk and uncertainty described above, there is a requirement for an array of skills that include plant and animal husbandry, knowledge of soil science, mechanical skills, ability with advanced technology, unique approaches to managing a multi-million-dollar enterprise, and the like. Because of the need to work extremely flexible hours often at very short notice and to multitask with the skills just mentioned, this part of the business model does not lend itself to conventional corporate management.

This is where the family farm, no matter what the size, depends on frequently unrecognized, multi-year apprenticeship of parents and grandparents, providing the necessary skills to sons and daughters. Farming, like very few other modern production activities, normally takes place where the family lives. For starters, all who live on the farm are constantly exposed to the ebb and flow of a host of activities along with the successes and failures of food production initiatives. Any sons or daughters who wish to continue as the next generation on the farm will be provided with guidance and direction over at least a decade, and probably longer, regarding all the necessary skills to operate the family farm.

My own youth was no different. By the time I was in high school, I could, without direct supervision, operate heavy farm equipment, shear sheep, deliver lambs when the ewe had birthing difficulties, milk cows by hand, administer medication and injections to ailing or injured animals, construct and repair fences, do basic welding, maintain farm equipment, and be trusted to feed numerous animals in an appropriate manner. In winter to keep the coyote population in check for the sheep, I ran a trap line, skinned the animals, and prepared the pelts for sale. With a shipment of up to thirty coyote pelts each season, I became a registered trapper with the Hudson's Bay Company, a historic four-century-old fur trading organization and the oldest company in North America.

To illustrate this point, the following schematic, from the website of the Canola Council of Canada,[19] captures the dynamic of decision making and the accumulated skills required when deciding what variety of crop to plant. This is only the business management element of farming and does not touch the many skills associated with actually planting, nurturing, and harvesting a particular crop.

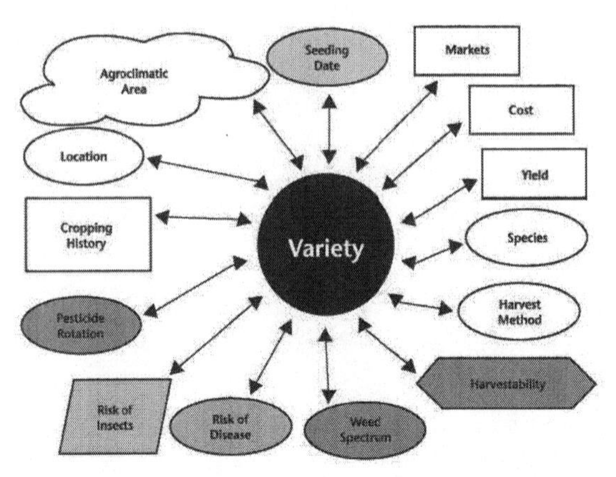

FACTORS INFLUENCING A FARMER'S SELECTION OF A CROP VARIETY TO PLANT

This array of acquired skills is typical for anyone growing up on a family farm. With this apprenticeship, when a young farmer reaches maturity, he or she has a substantial advantage over someone who wishes to enter the food production business as an adult. Successful late entry into farming is certainly possible through trial and error and adult education, but it is challenging, particularly for larger operations.

The typical farmer of today has an acquired skill set and an aptitude for dealing with risk and uncertainly often uniquely relevant to the farm on which he or she was raised. This equips farmers to take on the challenges of farming in a way that is generally elusive to today's urban-based corporate culture. Thus for those who are advocating smaller and more farms, some thought should be given as from where experienced farmers will come. To summarize, farming is not an easy vocation to acquire without the informal but effective apprenticeship found only in the family farm setting.

Regarding my current situation, while I was an active farmer with my father into my early adulthood, studied agriculture for six years at university and kept professionally in touch with the industry since then, I fully realize that I am not in a position to own and operate a commercial farm in an effective manner. Yes, I have the ability to produce some food, but to do so profitably and also achieve the full production potential of the land is not a likely outcome. I am not fully conversant with many of the advanced technologies and practices that had been introduced to farming over the past few decades. In my case, I am now in a position to at least know what I do not know when it comes to serious commercial farming. While novice farmers can make a go of small-scale farming, make some money, and certainly enjoy the fruits of their labors, the majority of large-scale family farms will only be effectively managed by those who grew up in the industry.

To illustrate this essential dynamic of farmers in food production in a very different place and time, in the early days of the Soviet Union Stalin killed or banished to Siberia the millions of Kulaks who were the productive Russian farmers prior to the Bolshevik Revolution of 1917. Unfortunately for this farmer class, they held out for a free enterprise system where a farmer should be directly rewarded for their efforts and not put on a wage and work on their erstwhile farms, which had been expropriated by the state and converted to a collective food producing operation. This collectivization of the farmland into industrial-type work units proved to be the fatal flaw in Soviet agriculture and perhaps the biggest weakness to the entire communist system. The assumption that was made by Stalin and his communist buddies was that all comrades were equal to any task that involved hard work, and that farming was just another productive activity. The model that they inflicted on the countryside had the routine of a factory, and thus the workers had no reason or incentive to respond to opportunities and challenges that are part of an individual farmer's daily life.

A fundamental example of the folly with the factory approach was that at the end of a working day at harvest time, operations ceased even if rain was expected the next day, which would both delay further harvest for perhaps a week and reduce the quality and quantity of the grain. If late in the season, even the prospect of snow the next day apparently meant nothing to the farm worker who collected the same few kopecks as long as he put in the prescribed hours of work. They did not appreciate the old proverb "Make hay while the sun shines."

The Soviet Union never came close to meeting their agricultural potential, and thus food security was a constant threat given the dependency on the likes of the United States and Canada during the Cold War to make up the food deficit through massive imports of wheat and other food products. This was ideologically a very bitter pill for Stalin and his successors to swallow, and it certainly was a factor regarding national security. Just think about it: without farmers, the Soviet Union had to rely on their Cold War adversaries to supply the most basic of necessities that the great communist system could not provide! This was in spite of having double the land mass of the United States and a little over half the population. I sometimes wonder why this undisputable fact was seldom made a major issue of by the United States during the Cold War, when both sides were touting their respective economic systems as being superior. The following bit of humor captures the abysmal failure of the communist system to deliver on feeding the population of Soviet Russia.

An American, a Czech, and a Russian were debating whose country had superior agriculture. The American says, "In America, we have developed such an amazing potato plant that within two months after planting, you can harvest it."

Not very impressive," says the Czech. "In Czechoslovakia, we have developed a potato plant that can be harvested only one month after you plant it."

"That's nothing," says the Russian. "Our potatoes are harvested the day after they are planted."

"That's impossible," say the American and the Czech.

You don't know hunger," says the Russian.

Late in the history of Soviet Russia, when Gorbachev was the minister of agriculture, he spent considerable time in the Canadian prairies because it is very similar to the agricultural part of Siberia as far as climate and soil are concerned. The soon to be premier wanted firsthand exposure to observe how a non-communist country could grow crops in abundance when harvests were not nearly as prolific in similarly endowed USSR. According to my Foreign Service colleagues who accompanied the delegation, Gorbachev was taken on a tour that included a visit to a typical supermarket. He immediately dismissed the event as a propaganda ploy, which the Soviets had often used themselves. He was convinced that his hosts had arranged a lavish display of quality foods in apparent abundance to impress foreign visitors. His Canadian handlers advised that such outlets were common, and during his forthcoming travels it would be his call where and when to repeat such a visit. Some days later he chose another supermarket at random and was apparently very much shaken to see the same wide array of fresh and high quality produce, meat, dairy, and other products in an equally attractive supermarket to the one he first observed and ridiculed as a phony.

He then could not help but to realize that that the supposedly down-trodden, ordinary workers of Canada had an incredible choice and abundance of foods and a superior diet than enjoyed by their Soviet counterparts—and all this in the North American equivalent of Siberia. Who knows how much this revelation impacted Gorbachev's subsequent thinking as he introduced Perestroika and other economic and political reforms that eventually led to the dismantling of the communist system in the USSR.

I suggest those who maintain that the present food system is broken and who have a distain for the supermarket reflect on what must have gone through Gorbachev's mind when he observed the abundance and choice of food that we take for granted.

Immediately after the Berlin wall fell, when I was with Agriculture Canada, I participated in an international conference in Budapest on the future of Eastern Europe and Russian agriculture. The recurring theme was that although the food production potential existed with such a large and rich land base, the outlook, particularly in the former Soviet Union, was that the absence of conventional farmers would seriously inhibit agricultural productivity, perhaps for decades. The situation was a bit brighter for the former USSR satellite states such as Poland, Czechoslovakia, and Hungary, where collective farms were not quite so vigorously forced upon the rural population, and many of the older country residents were farmers from an earlier era. In my report back to interested government departments, I entitled my synopsis of the conference "Where Have All the Farmers Gone?" to make the point that this very fundamental ingredient of highly productive agriculture was not there. Interestingly, two decades after the fall of communism, the Russian Republic minus the breadbasket Ukraine is now an exporter of wheat to the world because they have put farming back in the hands of farmers.

I have taken this historic and geographic diversion to illustrate the importance and uniqueness of farmers and their incredible work ethic and range of abilities, which remain the defining characteristic that cannot be found on a regimented factory farm for most types of agriculture. Thus so few US farms are operated with a conventional industrial production model.

Returning to North American agriculture, food production activities such as poultry (both meat and eggs), the later aspects of raising beef cattle, pork, and some dairy tend to be indoors or in a confined space without climate concerns, and they are routine and more in line with a conventional factory than land-based farming. Here the factory farm concept works quite efficiently and provides large amounts of quality and low-cost meat, eggs, and dairy. As a rule of thumb for consumers concerned about the factory farm concept, a significant proportion of animal proteins fall under this category while most food from plants are produced on family farms.

The fact that much of US and Canadian food comes from large family farms should not necessarily be seen as negative, but rather it should enable consumers to benefit from highly productive economies of scale and thus lower food prices. The fact that only some 2 percent of the national labor force[20] in both countries is engaged in primary food production is a sign of a highly efficient industry. Indeed, there is a very direct and real global correlation between a country's per capita income and the proportion of the population engaged in basic food production. Both the United States and Canada, along with Australia and a few others, fare particularly well by these standards.

I trust that the previous pages paint a promising picture of a dynamic rural America where the traditional farmer and his or her family remain the cornerstone of food production. Though the scale of operation for many may have expanded substantially, and a host of advanced technologies are in play, there is nothing seriously wrong with the current business model relating to the production of most of our food.

Even so, some purists would like to turn back the clock by promoting small farms that are diversified to retain the legacy characteristic of food production. This would mean that the number of farmers, if they could be found, would increase substantially, with each entitled to a decent standard of living. Given the drop in the economies of scale of smaller farms, this would mean that the cost of food would need to rise dramatically. Are consumers prepared for this scenario, even if it was possible? The acceptance of higher food prices without a guarantee that the quality and safety would be enhanced is very doubtful. The parallel of encouraging reduced productivity in other industries would be totally unthinkable. Take automobile production and subject it to the same model of legacy practices. For example, it is quite ridiculous to call for the elimination of robots on the production line so cars would be considered more the product of busy human hands, with higher priced cars and no quality advantages. So too is it with agriculture and the current large-family farm concept of producing much of our food.

On these large family farms, where there exists an impressive infrastructure of machinery, buildings, and other capital assets, there is little opportunity for farming to return to traditional operations for most of its agricultural production. These units are as essential to a highly productive North American agriculture as the Kulaks were to pre-Soviet agriculture. Certainly there is a role for smaller farms to play if the product is one (such as vegetables) that can be sold directly to nearby urban customers at retail prices. As stated previously, if for lifestyle reasons an individual wishes to be a part-time farmer and hold down other employment, this is a noble endeavor. However, as has been stated previously, it must not be overlooked that such small farming units provide only a tiny proportion of the total national food requirements to meet both domestic demand and export obligations.

As the saying goes, if it ain't broke, why fix it? We take an abundance of food at low prices in comparison to earning power for granted. Thus those who rally against the present system and call for a different structure in agriculture do so at considerable risk. The combination of skills, technology, land resources, and infrastructure is already in place, supporting this large, complex, and diversified industry that delivers one of the essentials of life. Serious tinkering with the structure of today's agriculture could have major unintended consequences regarding food security and they way we live.

5.

From Farm to Consumer

It may surprise some that the farm gate prices for many products constitute only a very modest percentage of the final cost to consumers. Take wheat, for instance, which is typically worth around five dollars per bushel to the farmer. One bushel weighing a standard sixty pounds is sufficient to produce ninety one-pound loaves of whole wheat bread. For white bread, with some of the more nutritive part of the wheat kernel removed, it is a little over sixty loaves.[21] In other words, for every loaf of whole wheat bread, the farmer's share is five and a half cents, and for the white version it is a little over eight cents. The same situation also applies for breakfast cereals and a host of other products. Even if the farmer gave the wheat away, the consumer would hardly notice.

Though the farmer would clearly like a better price, most can make an acceptable return on labor and other inputs with such an insignificant share of the consumer's food dollar thanks to the efficiency and rather large scale of modern farming. For both family farms and the modest number of factory farms, those involved in basic food production cannot be accused of becoming rich at the expense of the consuming public.

Many consumers have at least a basic knowledge of how food moves from the farm to their local supermarket, but they may not appreciate the overall complexity and scale of this activity. First there is the transport from the farm to the processing center. To many, transportation is a wasted effort that unnecessarily creates an excessive carbon footprint, and it would be much better if food production could be carried out closer to consumers. This makes sense wherever possible, but large populations and extensive fertile regions with superb farming potential seldom are conveniently located close to each other.

For example, take wheat and Kansas, the leading US state for this commodity, which regularly produces over 300 million bushels a year.[22] Given that a bushel of wheat is sufficient for ninety loaves of whole wheat bread, and the population of Kansas is approaching three million, the state produces enough wheat to provide each resident with nine thousand loaves of bread per year. New York State, with a population of around a twenty million and with somewhat less advantageous wheat growing conditions, produces only about seven million bushels a year.[23] When using the same basis of calculation, this comes to only thirteen loaves per inhabitant per year. It would appear that bread is a challenged product for New York adherents of the hundred-mile diet, except for those who make the effort to find a local supplier.

One might argue that wheat farming should shift to be nearer New York City, but this would be putting the proverbial square peg in a round hole. Kansas is the leading US state in wheat production for a good reason; it has extensive areas of quality soil and a climate that is ideal for wheat and it can produce it in abundance with a minimum of input compared to many places outside of the Great Plains. New York State has many positive agricultural attributes, but wheat production is generally one of the less profitable crops.

It is not only less costly in terms of energy and other inputs to produce wheat in most Kansas settings, but also consider that New York is better at producing apples than their western counterpart. Thus given the relatively low energy requirements to move such products by train, the grain or flour transported could easily be less of a carbon footprint for Kansas-produced wheat consumed in New York compared to the locally grown alternative. Also, Kansas wheat tends to have a higher protein content than the New York State version[24] and is therefore nutritionally advantaged.

Railroads advise that they can ship a ton of freight a total of five hundred miles on a gallon of diesel fuel.[25] Given that Manhattan, Kansas is about 1,250 miles from its eastern namesake, it takes two and a half gallons of fuel to move a ton of wheat, which is the equivalent of thirty-three bushels or close to three thousand loaves of bread. Thus a gallon of diesel fuel moves over a thousand loaf equivalents of Kansas wheat to the New York market. Food miles in this scenario are essentially irrelevant when you consider that driving a car twenty miles on a gallon of gas has much the same carbon footprint as providing a New Yorker with a loaf of bread from Kansas wheat at one per week for over twenty years.

By the way, baking is almost always carried out near consumption, given the relatively short shelf life of bread and the bulky nature of the product, which inhibits efficient long-distance transport. Thus in most large cities bread, but often not the wheat, is usually local.

I will cover more about the transportation carbon footprint in the chapter on the hundred-mile diet, but the point here is that transportation is an essential and acceptable activity to move a lot of our food from where it is produced to the place of consumption, and concern over food miles may not be a real issue. This is not to say that locally grown food is not an outstanding choice, and consumers should be encouraged to put a little effort into sourcing from nearby farms, particularly in the growing season for fresh fruits and vegetables. It is just a fact of mass urbanization that in our modern society, we are no longer required to live near our sources of food. This provides wonderful freedom of choice where we wish to live—and many may not appreciate it is dependent on the economical and efficient transportation of food.

Accepting that some transportation over distance is a necessity, I will introduce some instances where the basic commodities are delivered to food processing facilities, which, to take advantage of economies of scale, are likely very large and located where there is an ideal balance between distance from both the markets and the source of supply. For example, Heinz in Leamington, Ontario, with 80 percent of the total Canadian catsup market, produces 31 million bottles every year, using more than 250,000 tons of tomatoes grow locally.[26] Here the model is to process everything very near the source of the tomatoes and then distribute the product across a lot of geography. Leamington has an ideal climate for tomatoes and is located near the most densely populated part of Canada. Yes, tomatoes can be grown with varying yields in many locations, but what would be the advantage of many small catsup plants near each large population center?

Another example of concentrated food production is almond farming, of which over 75 percent of the world's supply is produced and processed in central California by approximately 6,000 growers.[27] Again, processing is near the source of supply but with little regard to the distance to markets. This production dynamic is due to a specialty crop being ideally suited to a unique climate, along with grower know-how and an established large-scale almond harvesting and processing infrastructure. Farmers everywhere are free to grow almonds, but can they do so as well as their California counterparts? Apparently this is not the case. The beneficiaries of this concentrated abundance are consumers everywhere able to get a reasonably priced, high-quality, and largely unprocessed nutritious food.

Another crop that is extensively grown in concentration in a well-defined region, and is even more remote from major population centers, are potatoes, of which Idaho accounts for nearly one-third of US production.[28] This concentration is particularly intense in the southeast corner of the state. Here processing is local but carried out in many dozen different facilities, from smaller bagging plants to very large fry manufacturing facilities. As it happens, to produce disease-free potatoes, grain crops must be grown the year after a potato harvest. In southeast Idaho, along with wheat, malting barley proved to be an ideal rotation crop, and thus Idaho Falls and the surrounding area is a US center for the malting industry for beer production in the United States and Mexico, for Anheuser-Busch and Grupo Modelo (and their Corona brand). All this takes place in a delightful and highly productive food-producing corner of the United States, far from any large population centers.

The primary production of food at the farm level is essentially based on climate, soil conditions, and the availability of large tracts of suitable land. Take food processing, which can vary from as small an effort as placing unshelled almonds in a bag to massive operations converting tomatoes into bottles of catsup. Such operations, because of transportation efficiencies and economies of scale, usually tend to be near the farms where the product is grown. There are notable exceptions to this, such as with baking, which has been mentioned previously.

Beyond these examples, what are the inefficiencies of food production with the presumed unnecessarily large carbon footprint due to transportation, and also a loss of freshness due to the time moving the product to population centers? Though there may be an element of truth in this in certain instances, transportation is but one fraction, and usually a small one, of the energy needed from production to the actual cooking of the food. By some estimates including my wheat example, transportation accounts for only 10 percent or less of the total carbon footprint of the food from production on the farm to the fork[29]. Economics of scale and good farming methods dictate that production usually takes place a long way from markets, particularly when there are such large and geographically and climactically diverse countries such as Canada and the United States with their highly concentrated population centers.

As one will expect for the folks in Kansas, California, and Idaho, the only bread, almonds, or potatoes respectively that they consume are local. Furthermore, all wheat grown in New York State is similarly milled, baked, and consumed reasonably near the farms where it is produced. I raise this point on where local food consumption takes place because one occasionally reads about the unnecessary and supposedly significant movement of food with long-haul trucks with the identical cargo of food passing on the highway, each destined to the starting point of the other. I suspect that this is more an urban myth than a widespread reality. Except for up-market items such as wine, cheese, and some high quality cuts of meat, old-fashioned economics dictate that for most basic food items it simply makes commercial sense to first supply the market nearest the point of production. This works because those in the food business—be it farmers, food distributers, or processors—have every reason to keep costs down wherever possible, and that very much includes transportation. Only if farmers and food processors cannot meet all local market requirements will food products move in from further afield where it is produced in abundance.

Moving closer to the fork, there is also much well placed concern that food processing, usually by large multinational corporations, destroys much of the nutritional value of the food products utilized to produce many of the products that we encounter on supermarket shelves. For example, catsup is a poor nutritional alternative to a fresh tomato. Before one blames the food processors as the sole culprits for such food degradation, know that they are only responding to what consumers are demanding, and here lies a major part of the very real issue of nutritionally challenged, processed foods. It goes back to the kitchen.

The cover article of the January 2, 2010, issue of the British newsmagazine *The Economist* features Rosy the Riveter[30] of World War II notoriety has a significant historic connection with the structure of our modern food industry.

The related editorial highlights perhaps the most important and positive socioeconomic phenomenon of advanced economies of the past century. The cover caption, "We Did It," refers to the fact that in countries such as the United States and Canada, sometime in 2010 women moved past sharing 50 percent of the labor force with men. This is in all respects a positive development, but one consequence here is less time for any family member to spend in the kitchen preparing meals from basic ingredients. Thus with substantially increased family incomes and a need to streamline the home life, the food processing industry stepped in and offered an immense array of convenient prepared foods.

One of the early commercial responses to this phenomenon was TV dinners, which were designed to replicate home cooking and often featured meat, a helping of potatoes, gravy, some vegetables, and perhaps a dessert. These became the subject of some ridicule because they tasted suspiciously like leftovers. The food industry appreciated that if they could not replicate the same food as prepared over a kitchen stove, they must come out with attractive and novel alternatives. It was well-known that the fundamental ingredients of salt, sugar, and fats in an array of combinations could please the taste buds in ways that the imaginative but nutritious-conscious mother who was preparing a meal in the kitchen would not provide except as a special treat.

By default, home-prepared foods were generally nutritious because the basic ingredients were mostly fresh and wholesome, and in most instances they were reasonably close to what left the farm. Though these same unprepared foods remain readily available and are generally at a reasonable cost in any supermarket, the tendency is to purchase convenience foods—or bypass the grocery store altogether in favor of takeout burgers, chips, and the ubiquitous sugar-laced soft drink.

Of particular importance today is the fact that today's food processing industry may start out with the identical products produced by farmers as homemakers used in their kitchens, the ingredients often deviate into so-called junk foods where nutrition was almost universally not a consideration. Flavor was everything with each new product subject to taste tests, and the best tasting product made it to the grocery shelf or fast food outlet.

When I worked for Agriculture Canada two decades ago, there was a section in the headquarters building that had an industrial kitchen staffed by qualified chefs whose focus was on developing new recipes utilizing the main ingredients coming off the land. As part of the routine, they frequently asked other employees to be their guinea pigs and participate in taste panels. I took part in this rather pleasant exercise on several occasions, and in looking back now, I realize that there were lots of cakes, cookies, cheese dishes, and hamburger variations—but nothing that I can recall emphasizing whole grains, fresh fruits, and vegetables or products that looked like they came directly from the farm. At that time the emphasis seemed to be on developing products for the food processing industry or providing homemakers with an easy way to replicate commercial junk foods. Incidentally my then employer has a new moniker, "Agriculture and Agri-Food Canada," reflecting an increased emphasis on nutrition. I suspect that a similar transition has taken place in the US Department of Agriculture.

In the kitchen of bygone days to where the time between cooking and consumption was a matter of minutes and food preservation was not an issue, preservatives were unnecessary and not even available to the cook in the kitchen; if food needed to be stored, it was in a dry form, canned, kept in a cool environment or frozen. What was a relatively easy kitchen effort was a major challenge for the food processing industry. Maintaining a long shelf life at room temperature became crucial if the food was to retain some semblance of freshness and avoid outright spoilage. In response a host of preservatives were introduced so that packaged processed food products remained unchanged from the time of manufacture through the wholesale and retail process and on to extended storage time in the home. Though the science is less than fully transparent on what these preservatives do in the body, there is some logic in the concept that any chemical that kills pathogens in food is probably not good for you. As a rule of thumb, the more ingredients listed on the food package that the consumer does not recognize as a food, the less healthy it probably is.

Unfortunately, in the preparation and marketing of such foods, nutrition largely fell by the way, and consumers were often left with empty calories and food that stored well, but only thanks to the preservatives with their questionable impact on health. This is not a good state of affairs, but to rally against the food companies is misdirected criticism. Savvy consumers and their shopping habits have the greatest impact on improving the quality of processed foods.

Many food processors and fast food providers are not serving our nutritional needs particularly well, but they are only responding to a market dynamic by providing what the consumer has been asking for, which was convenience, shelf life, and flavor at a reasonable price. Unfortunately, until relatively recently both processors and consumers paid scant attention to nutrition, and today we have obesity, diabetes, and other health issues directly attributed to this sad state of our grocery basket affairs. Fortunately there is an increasing critical mass of the population realizing that the quality of food matters a lot and that something needs to be done.

In response to this, organic foods are gaining in popularity, there is the local fresh food movement, we are observing untold studies and publicity regarding this issue, and there is pressure on our elected officials, from school boards to law makers in capitols, to make nutrition an issue of the utmost importance to offset this very real health epidemic.

This is not going unnoticed by the food processing industry. However, it will be the critical mass of consumers choosing the healthy foods over the junk varieties that will encourage food retailers to push processors to make a sustained and large-scale effort to put nutrition back into foods. There are signs this is happening, with the US Grocery Manufacturers Association[31] maintaining that they now have ten thousand different manufactured food items with reduced calories and less sodium and sugar. Indeed, one can observe many products on the supermarket shelf that claim to have reduced one or more of these factors. One example illustrates this point. I have found two brands of peanut butter in local grocery outlets that have but one ingredient, peanuts—with no salt, sugar, or preservatives.

Moving away from food processing and further along the food chain from farm to fork, whether food is fresh or processed, it usually goes to large warehouses, with each grocery chain having a central food distribution depot in all of the larger population centers. This is a necessary staging area that brings together many of the thousands of products one finds in a supermarket. The last stage is delivery by truck to all outlets, be they a monster box store or a corner grocery. Nothing fancy here, but this step ensures just-in-time delivery so that shelves are always full when we show up to do our shopping.

In spite of the success of the mostly seasonal farmers' market, it is through the supermarket that the vast majority of food reaches consumers. According to the Food Marketing Institute,[32] there are over thirty-four thousand supermarkets in the United States, each with an annual turnover exceeding two million dollars. In 2006 the average number of items offered and sitting on the shelves in coolers and freezers was approximately 45,000, and by 2009 this had grown to 48,750 items. The following table captures some interesting statistics regarding food retailing industry.

Selected Statistics on Supermarket Operations	
Number of employees	3.4 million
Total supermarket sales—2009	$556.973 billion
Number of supermarkets—2009 ($2 million or more in annual sales)	35,612
Net profit after taxes—2009	1.22%
Median Total Store Size in Square Feet-2009	46,235
Median weekly sales per supermarket 2009	$485,346
Percentage of disposable income spent on food—USDA figure for 2008 food-at-home food away-from-home	5.7% 4.0%
Weekly sales per square foot of selling area—2009	$11.77
Sales per customer transaction—2009	$29.24
Sales per labor hour—2009	$158.68
Average number of trips per week consumers make to the supermarket—2009	2.1
Average number items carried in a supermarket in 2009	48,750

Source: Food Marketing Institute

According to the same source, the average profit on turnover was an amazingly low 1.22 percent after taxes. It is all about volume turnover, with the average consumer actually paying only about a dollar and a quarter out of a hundred-dollar grocery bill for the convenience of being able to secure all their food needs in a very short time at one location. Though clearly profit driven, the food retail industry is extremely competitive, with many players offering essentially the same options in most neighborhoods. Consumers should not concern themselves that a significant proportion of their grocery bill is going to the profits of the supermarket chains.

Another interesting statistic is that there are 1.5 employees in supermarkets for every farmer.[33] According to the above table there are 3.4 million supermarket workers; the US Bureau of Labor Statistics lists the number of farmers as 2.25 million.

Given that there are tens of thousands of items offered in a typical grocery store, those consumers who are health conscious should easily be able to pick and choose a host of products that provide both variety and nutrition when served at the family table. Also, if a greater proportion of the population becomes increasingly health conscious, this will result in additional products such as the natural, peanuts-only peanut butter that I mentioned becoming standard fare.

Over the past fifty years, there have been many food safety issues that have been effectively addressed. All processed food product are labeled with their shelf life and nutritional ingredients. The USDA Food Safety and Inspection Service[34] and the Canadian Food Inspection Agency[35] have small armies of qualified people with regulations and enforcement measures in place that are constantly involved with this issue. Furthermore the Food and Drug Authority[36] as well as Health Canada[37] also play very significant roles regarding food safety. This issue will be dealt with in detail in the chapter on food safety, but it is mentioned here as an important step in the process as food moves from farm to consumer.

Indeed the food handling and processing industry is in many ways self-policing and has the highest regard for food safety. This is not entirely altruism given the damage to reputation and the costs of administering a recall if a product enters the market and then is determined to be a health risk. In this regard the corporate bottom line is much better served over the long run if full compliance to food safety regulations and standards become a core corporate value. Although the food chain supply situation from farm gate to grocery checkout may be considerably less than perfect, it is far from broken, and in some ways it is amazing that an industry that is so complex, involving so many products from so many farms and processors that are consumed by hundreds of millions of people, works as well as it does.

The food industry is often accused of being only interested in profits, with too little regard to the consumer. Profits are the cornerstone of free enterprise, so this part of the accusation is quite correct, but why should this essential element of the economy not be profit driven? This economic system has its flaws, but it seems to be the best one out there that rewards financial risk, entrepreneurship, and hard work. For an industry as vital as food, it would seem folly to somehow expect or demand that the profit motive be set aside for the good of humanity. Thus when I read or hear critics of the industry lamenting about corporate greed, I cannot help but wonder, if investors and the talent behind feeding the world were somehow expected to contribute without the financial rewards expected elsewhere in the economy, would there be the vibrant industry that we have today? Most certainly not, and the unintended results would be devastating for all except the prosperous who, ironically, could handsomely reward and provide profits to those catering to consumers who could still comfortably afford whatever food costs (they can always shop for the very best at what I jokingly refer to as "the hundred dollar store") while those toward the other end of the economic spectrum go hungry.

As stated previously, because of the intense competition between food processors and also various food retailers, profit margins are slim, and the factor that makes the industry viable is the very high volumes involved. The high volumes and low profit per unit of output also apply to the commercial family farm. While boutique operations in the food industry have their very real appeal and can offer advantages in service, quality, and other lifestyle choices quite rightfully enjoyed by the prosperous who are prepared to pay the price, it is only the efficiencies and economies of scale that large-scale operations offer that can provide so much food for so little money, compared to the situation even half a century ago, when a much larger proportion of take-home pay went to feeding the family.

This same competitive dynamic between food processors can have an impact on the nutritional quality of the processed food products offered at the retail level. Unlike the tobacco industry, which was unable to modify its merchandise mix and maintain a large market presence, the food industry has no such concerns if products are deemed unhealthy, because people will always need to eat. Though some brands will definitely suffer, a reasonably nimble food processor will be able to profitably follow a market trend led by a nutritionally savvy public.

A handy rule of thumb on the selection of healthy foods is the more it looks like the product that leaves the farm, the healthier it probably is. Fresh chicken versus chicken nuggets is another very obvious example. Fresh, frozen, and canned peas are three variations, with the latter two moving increasingly away from the farm. This approach to food should not be too difficult for the typical shopper to manage.

Yes, the food processing industry could do more to bring us nutritious food, but to give *caveat emptor*—or "let the buyer beware"—a slightly different twist, I suggest, "Let the buyer be aware." A well-informed public making wise food purchases will do more to remove less than healthy food products from grocery shelves more effectively and faster than any other approach. Food enthusiasts and food writers are right to rally against what is offered. They can lay all the blame they want on the food industry, but little will happen without placing much of the responsibility in the hands of the consumer.

Take salt in prepared foods. There seems no ambiguity that high sodium intake has a negative impact on health, particularly cardiovascular diseases, and yet in spite of sodium content being clearly labeled on every package of prepared foods, most food processors still continue to place more salt in their products than most nutritionists would deem healthy. The actual activity of including salt in food processing is an insignificant step that can be adjusted without effort or cost, so why do most food processors continue to provide an excessive amount of salt in their products? Consumer preference is the obvious answer. If shoppers genuinely cared enough about this issue, low-sodium products would outsell the alternative. One could argue that if the food processor was behaving responsibly, they would remove the salt voluntarily. However, with apparent apathy on the part of their customers, why should they change the flavor of a product and run the risk of losing customers?

For better or worse, the customer is always right, and very much so when it comes to providing what they want in food. Again I emphasize that the poor nutritional aspects of many food products could be rectified by the consuming public making the right choices. And yet in an Australian study[38] it was determined that in spite of an extensive information campaign on the dangers of excessive salt intake, only an insignificant 1 percent of the disease risk in the general population was averted through consumers making a better informed choice in their food purchases.

Thus, in spite of a call for less involvement for government in our lives, if the food purchaser makes bad choices that impact their quality of life and the expenditures on health care, consideration must be given for regulations to mandate issues such as a reduction of the sodium content in food. As I will argue in the next chapter, perhaps consumers need to be protected from themselves.[6]

6.

The Consumer of Food

Thus far I have brought the food from the farm and now almost to the fork, where the consumer is totally in control of this last, rather lengthy, and complex dynamic. In my mind if any part of the process is broken, it is the part of this journey that is totally in the control of the consumer.

I shall start with the beverages. The average American consumes 150 quarts, or 450 twelve-ounce cans, of soda pop per year.[39] Assuming that fifty of these are of the no-calorie or diet variety (which have their own negative issues), and the remaining four hundred cans are the typical sweetened type (which usually has 160 calories of low-quality carbohydrates in the form of sugar), this means that sixty-four thousand calories—or the equivalent of nearly a month's worth of a typical person's food requirement—are ingested as "junk water" to accompany junk food and, to borrow a term usually associated with a lake or stream that fails to make a sports fisherman happy.

To look at this calorie intake in a different way, to lose a pound of weight, a person must either avoid the equivalent of 3,500 calories in their diet,[40] indulge in increased activity to burn off the weight, or more likely a combination of the two. This means that the soda pop consumed annually is equivalent to about eighteen pounds of body fat. These are empty calories with no associated nutritional benefit, and they only deliver a brief but positive flavor sensation. As is widely documented, this excessive consumption of soda pop represents a very poor lifestyle choice, particularly by those who consume in excess of the average!

Other than soda pop, there are a host of energy drinks, fruit cocktail concoctions, and other beverages that contain similar quantities of sugar-based calories and thus also qualify as junk water in spite of the fact that they might contain some fruit juice.

Unfortunately, beyond the informed minority, a large proportion of the population has little concept, and perhaps cares even less, about the harm they are doing to their bodies by the excessive consumption of sugar-laced beverages. For these folks the freedom of choice is paramount, and they may have a point, but better if it was an informed choice regarding the folly of their ways.

For any effective program on reducing sweetened beverage consumption, it must be first recognized that that beverage consumption and socialization often go hand in hand, and brand recognition has a lot to do with it. Although water in a glass from the kitchen tap is a perfectly good thirst quencher, it does not have much social appeal. Here is where I question the movement to ban or discourage the consumption of bottled water. Yes, the argument is sound that the product is essentially identical to tap water, except that it comes in a plastic bottle with environmental consequences. However, cans and bottles of soda pop get a free ride on this issue basically because it is different than the product available at the kitchen sink. The bottled water industry points out this double standard along with the observation that the availability of their product at all beverage points of purchase could play a meaningful role in reducing the high-calorie alternatives.

The absurdity of this crusade to avoid bottled water became apparent when I recently presented at a conference held at an up-market hotel in downtown Chicago. As this was the fourth time I was a speaker at events at this hotel, I knew the drill reasonably well as far as the type and style of catering. Until the most recent occasion, refreshment breaks included a variety of soft drinks, coffee, tea, and bottled water. This time there was the full array of beverages—but without the water. Given that the hotel cannot charge for basic ice water by the glass, none was offered. Thus if someone wanted a cold beverage, the unintended consequence of this movement against bottled water was that the choice was limited to soft drinks that were, of course, in disposable cans or bottles.

I will perhaps risk my credibility and make some recommendations on how the excessive consumption of high-calorie, low-nutrition beverages could possibly be reduced. My concept is to treat excessive consumption of junk water as a form of substance abuse given the obesity, diabetes, and other health issues that are the result of such behavior. Here are my ideas.

1. Follow the Norwegian example[41] and put a tax on such beverages, both to discourage consumption but to augment revenues to cover the additional tax-supported health costs incurred. I hear

the argument that this was contrary to basic taxation principles, because the poor would pay proportionately and perhaps even more in absolute terms than their generally more health-savvy, wealthier counterparts. It is surprising that this point of view has any legitimacy; the "sin" taxes on tobacco and alcohol have the same impact on the poor and yet exist as a reasonably effective means of reducing the extreme abuse of these substances in most jurisdictions. For those concerned about this apparent injustice, they seem to overlook that beverage companies, who are constantly in a competitive mode and seeking market share, will quickly introduce products that are healthier to avoid the tax to maintain an attractive price for their product and maintain the same level of market penetration as the overall liquid intake of consumers will not likely decrease.

2. Mandate prominent warning notices on such drinks. These could take such forms as "This product has no nutritional value and could lead to obesity, diabetes, etc." As is the case with cigarettes, the message could vary to include specific warnings about disease and excessive sugar consumption.

3. Encourage the consumption of healthy beverages, including low–calorie, pure fruit juices and, yes, bottled water. Perhaps a novel idea would be to mandate that half the containers in a case of canned or bottles sweetened beverages must contain the healthier variety. This would at least move a healthier product into homes that probably would get consumed in place of the beverage that should be avoided. For example, in our household we purchase canned low-sodium soda water by the case but no pop, so our teenage son has a somewhat socially acceptable beverage to offer his friends. It works as more and more of his friends are being made aware of our health-based reasons for such behavior.

4. Continue with the excellent movement to reduce the point-of-purchase opportunities such as vending machines in schools and other public institutions, particularly those locations frequented by younger members of our society who would not likely read any advice on the package.

These recommendations, if considered, will not be well received by the high-calorie, sugar-based packaged drink industry. They own valuable brands that are globally recognized, and for good business reasons they would prefer the status quo. Unlike the tobacco industry, which faces product elimination, the beverage manufactures can adapt their product and still remain in the marketplace, perhaps even using a version of their existing brands. After all, they are in the business to sell drinks and not necessarily just flavored, carbonated, sugar water.

For those who have concerns about issue of the "granny state" being overly concerned about our well being regarding what we ingest society has long accept the governments role on how we drive, drink, smoke, can display our body in public, treat our fellow citizens and a host of other matters that impact our daily lives. Why should something as important as how we eat and the benefits regarding quality of life and savings for the health system either through reduced government support and taxes or improved insurance premiums not dictate some form of government intervention.

To provide further insight on what might be done to promote healthy beverage consumption, the California Center for Public Health Advocacy[42] has posted these common-sense suggestions that mirror my own ideas.

CA Campaign for Healthy Beverages November 2010

Local Beverage Policies: Soda

1. *VENDING MACHINES. Eliminate the sale of sweetened beverages in vending machines on city or county owned property.*
2. *PUBLIC PROPERTY. Eliminate the sale of sweetened beverages in city or county owned property, or at any city or county sponsored event, meeting, or program.*
3. *SCHOOLS. Establish policies to eliminate electrolyte beverages in schools.*
4. *MARKETING AND SPONSORSHIPS. Eliminate marketing of sweetened beverages, including sponsorships of and the presence of logos in schools and at city or county sponsored programs or events.*
5. *YOUTH VENUES. Eliminate the sale and marketing of sweetened beverages at zoos, museums, parks and other places frequented by children.*
6. *CHILDCARE, AFTERSCHOOL SETTINGS. Eliminate the provision or sale of sweetened beverages in childcare and afterschool programs.*
7. *BREASTFEEDING. Ensure that breastfeeding is supported at workplaces and in public buildings/events.*
8. *PUBLIC FUNDS. Eliminate the purchase of sweetened beverages by a city or county.*
9. *CHECK-OUT LANES. Enact a city or county resolution encouraging retailers to remove sweetened beverages from check-out lanes.*
10. *SIGNAGE. Strengthen city and county signage ordinances to limit the amount*

and type of signage on stores and buildings. (The ordinance must apply to all products and all signs because legally it cannot target a single product type.)

11. *DENSITY OF RETAILERS. Limit the number and/or density of sweetened beverage retailers near schools and playgrounds.*
12. *RESTAURANT INCENTIVES. Establish nutrition standards for meals that include toy-giveaways and other incentives.*
13. *TAXES. Establish a city or county tax on sweetened beverages and use the funds to support local nutrition and physical activity efforts.*
14. *CORPORATE AND ORGANIZATIONAL POLICIES*
    a. *Vending: Eliminate the sale of sweetened beverages in vending machines.*
    b. *Water: Ensure the availability of free good tasting water.*
    c. *Marketing: Eliminate marketing of sweetened beverages, including sponsorships and the presence of logos.*
    d. *Purchase: Eliminate the purchase of sweetened beverages.*
    e. *Breastfeeding: Ensure that breastfeeding employees are supported.*

Besides beverages, much has been written about the extensive array of unhealthy food products that are so readily available and at low cost. I will make no attempt to repeat the material that some excellent writers have already covered; instead, I will take the potentially controversial next step and make some additional recommendations that might result in better food choices by the consuming public.

1. As with the recommendation regarding beverages, tax those products that clearly exceed dietary limits of fat, salt, or sugar. This will provide an effective incentive for the manufacturers of such products to provide an alternative product with more reasonable levels of these ingredients.

2. Mandate warning labels that are displayed prominently on the package about the nutritional qualities of the product. These could be worded along such lines as "Contains excessive amounts of sodium, fats, and/or sugar," or "Is of minimal nutritional value." This idea will meet substantial pushback from the food processing and packaging industry. However, such regulations will present an even playing field for all participants in the industry, and if introduced over a reasonable period of time, it is expected that valuable brands can be preserved, particularly as the consuming public will be acutely aware of why their taste buds have a different reaction to the contents of a familiar package.

3. For packaged food products that offer a nutritious choice such as whole wheat bread, enable the manufacturer to make a positive statement that is certified by some recognized third-party agency that the contents exceed some minimum standard.

4. The "Nutritive Facts" table on all existing packaged food products is difficult to fully interpret in the short time available at the point of purchase by many people, including me. To simplify matters, perhaps carry the previous idea to a grading of all foods and beverages according to their overall nutritive value. For illustrative purposes, I suggest a system that would award a zero to most brands of soda pop as they exist today, and a ten for truly nutritious products. This would also be controversial, but if the numeric rating was established on widely accepted nutritional standards, as determined by a reputable third party, perhaps it will work.

5. Revamp the "Nutritive Facts" table to be more precise on what constitutes a serving. Presently it can be 1 dozen potato chips, one-sixth of a package, a slice of bread, or some other measurement unique to the product, which makes comparisons very difficult. I propose that the nutritive facts pertain to a standard quarter-pound of the product, or one hundred grams. In addition to this, the manufacturer can post the corresponding facts for whatever they deem a reasonable serving to be. An Australian friend recently sent my son a care package including unique food products from that country. I noted that this was the practice adopted there, so this suggestion should not be too difficult to implement because the major brands there are much the same as in North America, and manufacturers would be familiar with such an initiative.

6. Forbid all claims of the nutritive qualities of the product except those sanctioned by some authority. At present these claims can be misleading or even an outright falsehood. For example, on a pretzel package, the contents of which contain obscene levels of salt, one could not help but notice a statement in rather large print that read "0 Trans Fat." Though this is a reflection of reality, it provides a false assurance to the casual purchaser that the product has health benefits when it clearly does not.

Currently food processors with packaged goods are encouraged to voluntarily improve the nutritional content in response to the masses of information demonstrating that a majority of their products fail to meet the dietary requirements of their customers. As previously stated, the US National Grocery Association claims that nutritional improvements have been made to some ten thousand products. While helpful, there is little evidence that the improvements made are indeed substantial across the board and apply to best-selling, well-known brands.

As with beverages, there is a good business reason to stay with the products consumers now willingly accept, and thus to avoid the risk of market share loss, the status quo is the preferred position. Indeed, it could be quite risky to change a product that currently sells well because of a consumer preference for the existing combination of fats, salt, and sugar when their competitors are not under any obligation to move with any urgency on the matter. Thus voluntary improvements are unlikely to contribute significantly to improved nutrition in processed foods.

Recently there was a government-sponsored campaign in Ontario to encourage reduced salt consumption. Instead of something so blunt as "Too much salt kills," the message was a fairly gentle reminder to watch sodium intake. Although salt is a very general chemical term, and sodium is a more precise scientific description of sodium chloride, or common table salt, the organizers of this campaign missed the opportunity to use common vernacular, and the message was probably lost to a lot of folks who are challenged regarding their nutritional expertise.

Thus public pressure is required to encourage the government to mandate such changes with clear nutritional goals, along with an understandable message to consumers. As such an approach will apply equally to all players in the industry and creates a level playing field where altering the product can proceed with the full knowledge that all competitors are in an identical situation. Unless the industry is forced to make the changes in their products' nutritional content, real progress in this regard will be slow or even nonexistent.

Perhaps my recommendations are naïve and need refinement, but clearly something must be made to move past the existing intolerable situation regarding available food choices. Every effort should be made to establish a food system where nutrition, and not exclusively taste, dictates the average consumer's choice on what food to purchase and consume.

What I am proposing is changing the dynamic of the market in favor of healthier foods, but in a way that provides the same product disruption for all food processors and an equal opportunity to move through the transition in a profitable manner. Freedom of choice should not be a factor for consumers; they will still have the option to purchase a great array of food products as before while manufacturers will still be equally free to win and maintain customer loyalty including altered existing brands.

As stated previously, although there is always opposition to changing the status quo, this is not like what befell the tobacco industry, which for good reason faced severe and effective regulatory restrictions to discourage smoking, and it could offer no alternative product opportunities close to the core business of the industry. Cigarettes were not banned, so those who oppose government interference with choice opportunities for individuals should not be unduly stressed. However, with health costs rising, perhaps those who refuse to follow a healthy lifestyle will not have the luxury of having their misdeeds covered by more responsible taxpayers and fellow health insurance policy holders.

It should not be overlooked that the amount of food purchased in quantitative and dollar terms can be expected to remain so that overall industry profit need not be diminished in spite of the fact that there will be winners and losers during the transition. If these suggestions or some version thereof are ever adopted. it will be interesting to see who the early food industry adapters are and what it might do to positive market share for them and their profits.

7.

Role of Government

Most governments strive for a well-functioning farm-to-fork dynamic that results in a secure, low-cost, and safe food supply. Some are doing a better job of this than others, but none are perfect. This chapter will focus on the government's role in maintaining an adequate supply of food, and subsequent chapters will dwell on the important aspects of food safety and the bookend factors of food security and insecurity.

Most consumers in North America accept food security as a given and take it for granted that safe and relatively inexpensive food will await them in abundance in grocery stores, farmers' markets, restaurants, and a host of other outlets. Few give a second thought to this, but perhaps they should reflect on this issue, particularly as food production is the riskiest of commercial activities, with weather being a perpetual wild card that can reduce a harvest to nil in a matter of an hour in the case of severe storms, or over weeks and even years if there is a prolonged drought.

In 2007 Australian wheat production was down significantly because of drought, which resulted in the global stocks of this grain falling to their lowest level in twenty-five years. This caused wheat prices to quickly double to over ten dollars per sixty-pound bushel and, in turn, impacted on the price of other basic crops such as corn and rice, which also saw dramatic prices increases in spite of record harvests for both crops. Indeed there never was a shortage of corn; the United States eventually had a reasonably healthy carryover at the end of that crop season of over 1.3 billion bushels[43] which is the amount of corn that is typically used for both human food and seed for the next year's crop.

However, even the possibility of global food shortages caused havoc when the basic food prices spiked, and several countries faced food riots and disruptive demonstrations, including India, Bangladesh, Yemen, Cameroon, Egypt, and especially Haiti where the government fell as a result of the unrest. Even in North America, consumers of items such as popcorn and bread noticed a price spike that did not fall particularly quickly after the commodity market returned to normal.

In this instance when only a concern about shortages, and not an actual shortfall, caused so much turmoil, what would happen if the global reserves of basic food commodities were actually depleted and could not be replenished before the next harvest? With little doubt the results would be catastrophic. Many nations would face severe civil disobedience, a disruption in commercial activities, unstable government, and an inability to provide effective law and order. The stability of otherwise reasonably well-governed countries would be threatened to the point where states might fail. Besides the human suffering, such a scenario is one that the United States, with its interest in global security and stability, would find extremely challenging.

Here is where the much maligned Federal Farm Safety Net Program[44] is essential as a form of insurance enabling farmers to plant and harvest crops with some of the risk shared by the government. It may appear to the taxpayer that they are providing aid to an industry that seems to be prosperous on its own, but the alternative would be for risk-adverse farmers to limit their inputs, and thus yields, whenever conditions such as low agriculture commodity prices or adverse weather were a factor. Although there are abuses to any system that involve massive government funding to a very large number of recipients, the fundamentals of the safety net transfer payments is a sound one and is by no means unique to the United States and Canada.

The European Union is in many ways even more generous with farmers with their Common Agricultural Policy.[45] Take Germany, where the older generation still recalls the extreme hunger many encountered at the end of the Second World War. Given this background and the fact that the country has over eighty million inhabitants—and yet it is slightly smaller than Montana—food security is an issue of national importance with a higher level of public awareness than in Canada and the United States. Thus to ensure a robust farming community constantly producing ample food, governments must share the risk and ensure a reasonable level of farmer prosperity with taxpayer dollars.

But why should taxpayers be on the hook for such support when there is a large and effective insurance industry that seems to be in a position to insure almost everything? For some specific threats to harvest, such as hail, where the risks can be calculated, insurance is available because the industry has determined that in any given year only a small percentage of farmers are affected. The situation changes dramatically if farmers seek insurance covering all risks to yields and to commodity prices below the cost of production. Given the massive scale of the farming industry and the uncertainty of bundling all such risks, the insurance industry gives such coverage a pass because it is beyond their capacity to meet the commitments of a catastrophic year with wide, multistate, adverse growing conditions or a collapse in the prices of all major crops and livestock.

One may also ask why then do governments seemingly go overboard and set programs that constantly appear to produce surpluses. Given the vagaries of weather, it is not possible to fine tune assistance to farmers so that supply is close to demand. Returning to the example of wheat and Australia and the havoc that one country's crop failure had throughout the world, food shortages are much more important to avoid than surpluses. Hence a successful government program in support of agriculture must err constantly on the side of oversupply. In the United States and Canada, the generally bountiful harvests go a long way in international trade to providing food, most often on a commercial basis, to countries with a population base that exceeds their own ability to produce food. Global stability is enhanced, and otherwise empty bellies are full thanks to government intervention to insure that food supplies are adequate on a global basis.

Actually the concept of food surplus is a misnomer, as some are led to believe that it is akin to army surplus, where it is no longer needed for the original intended purpose and must be disposed of or even discarded. Think of the global food supply at any one time as a parallel but a vastly larger variation of the food stored in the fridge, freezer, and kitchen cupboards of a typical middle-class home. Though the stored food is "surplus" to any one day's family requirements, most if not all is eventually consumed. Furthermore, just as the seasonal harvests in both northern and southern climes, plus the perpetual harvest in the tropics, continuously contributes to the global food supply as populations eat their way through earlier deliveries, households replenish the food for the family at about the same rate as meals are served at the table. It is this balance that responsible governments strive to achieve. Without a buffer of three month's supply of food, the FAO considers the world to be in a precarious food security situation.

Given that the US Farm Safety Net Programs cost considerably less than 1 percent of the federal budget, is this not a modest price to pay to ensure that food security is not a significant concern, both domestically and with friendly countries that lack a bountiful agriculture sector? Even the wealthy that would not likely face personal food security issues would certainly be impacted if hunger became an issue to a significant proportion of the population, both at home and abroad.

Some complain that national, state/provincial, and local governments are lackluster in their support of popular initiatives such as the organic movement, locally produced foods, and even urban chicken production. I am not privy to exactly why the support for these movements is not as great as some might like, but as I will discuss elsewhere in this book, the overall actual and potential impact on national food output by these movements is not significant. Perhaps elected officials and bureaucrats are of the opinion that these positive movements are generally successfully supported by the relatively prosperous, and the government should focus their vigilance on overall food supply.

Another concept of government is that there is a three-way conspiracy including elected officials and bureaucrats, commercial agriculture (which is mostly made up of larger family farms), and the agriculture technology providers (which include the supply GMO varieties and chemicals). I wish to assure readers that there is no conspiracy in any negative sense, but the three must work together to maintain an ample and reliable flow of food for domestic consumption and exports.

The government also plays a major role in the area of food safety which will be covered in detail in the chapter on this topic.

In summary, by far the most important role of government in agriculture is to ensure that there is a robust, efficient, and as sustainable as possible large-sale farming industry that has the best possible opportunity to provide ample, high-quality food, and to do so without disruption or worry about shortages and widespread food insecurity.

8.

Farming, the Environment, and Biodiversity

While the symbolic fork may not be an apparent element of this chapter, the environment and sustainable farming practices do impact how much food will be available to consumers in the long run. Also, compromises on land use for non-food production have an impact on the farm-to-fork dynamic.

Farming is arguably the most destructive environmental activity of mankind, given that billions of acres are impacted by the need to feed the global population. But as Winston Churchill once said in the British House of Commons, "Democracy is the worst form of government, except all those other forms that have been tried from time to time." So it is with farming—there is no apparent alternative but to disrupt nature to feed the world. The overriding objective is to minimize the disruption while still meeting this very real food need, so there will be environmental compromises that must be offset, wherever practical, with mitigation practices on the part of farmers.

Almost invariably farmers tend to be good stewards of the land, if for no other reason than they wish to maintain the value of their real estate asset and to be assured of good yields in future years. Furthermore, why would they wish to live and raise their families in a setting that was unnecessarily environmentally challenged? These points are worth dwelling on because some contemporary environmentalists, journalists, and writers portray quite the opposite.

The terrible depredation of farmland that occurred prior to World War II because of poor farming methods gave rise to the term "dirty thirties," when wind erosion took a devastating toll on vast areas of North American farmland. Much was learned from this, and practices such as minimum or no tillage and leaving sufficient residue on top of the soil to minimize wind and water erosion are nearly universal in their application. Plows that buried most or all the residue from the previous year's crop have become obsolete and are replaced by more environmentally friendly means for planting and growing crops. It is now a mandated practice in many jurisdictions, not only in the United States but elsewhere, for natural drainage waterways to be planted to grass, which drastically reduces erosion from runoff.

It may surprise readers, but extensive water erosion is a fact of life that predates farming. The rich soil of the Mississippi Delta originated thousands of miles upstream from as far away as southern Alberta; it was deposited there over many thousands of years before the first farmer set foot in North America. To be certain, the presence of agriculture exacerbated the situation, but it should not be overlooked that erosion is also a natural phenomena as well as a man-made environmental concern.

Yes, there are millions of farmers, and all are not perfect, with a small percentage practicing poor soil management. These are by far the exception, and though their misdeeds often achieve public attention, one should not write off the role farmers play in both maintaining their soil but also the millions of acres of wildlife habitat that they provide on their non-productive land.

This leads me to the subject of biodiversity, or the lack thereof, which many critics of commercial farming use as evidence that larger farms and their extensive fields of a single crop or monoculture are undesirable. As stated at the outset of this chapter, food production by its very nature is disruptive to the natural flora and fauna that once thrived undisturbed on virtually every piece of land before the European settlers arrived in North America. Unfortunately, the trade-off is food for people and less biodiversity.

Basically there are two separate issues relating to biodiversity. The first is that for extensive monocultures such as corn production, the potential for disease or pest infestation is increased compared to farming practices with several crops grown on a rotational basis in the same area. This is a lesser issue to environmentalists than to agriculturalists and others concerned with stable food supplies required to feed all of humanity.

The second issue relates to the lack of opportunity for biodiversity when single crops dominate so much of the land in a particular area and thus provides limited opportunity for other plants and animals to thrive.

In keeping with corn as an example, perhaps there is no greater concentrated monoculture anywhere than is practiced in the Corn Belt of Iowa, Illinois, Indiana, Minnesota, and neighboring states. Corn, and for that matter also soybeans, grow spectacularly well in this region and provide the maximum returns both in income and pounds of food per acre. Other crops such as wheat can be produced here, but financial returns and the overall food production would be less. Meanwhile, wheat does well elsewhere, where the season is too short or there is too little rainfall for corn.

The economic advantages of concentrating corn production—that are supported by advanced agricultural technologies that mitigate insect, fungus, and weed infestations—seem to make the monoculture risk acceptable. Furthermore, when the call goes out for greater crop diversity, the typical corn farmer would not likely have the equipment, the appropriate scale or type of storage facilities, or the know-how to effectively produce other crops given that he or she has focused on growing corn and perhaps soybeans for many years.

Returning to the second concern over monocultures and the lack of biodiversity, this is the very real concern regarding the near absence of any other plants and a limited variety and population of animals, birds, and insects in a field dedicated to any one crop. In my opinion, when a field of any size, even as small as an acre, is planted to an annual row crop, there is a loss of biodiversity, so smaller fields with a greater diversity crops but on the same number of acres does not substantially alter the environmental impact of a highly productive farming region.

However, I ask readers to judge for themselves about the biodiversity that actually exists throughout the continent and in particular the United States. America's land mass is approximately 19.5 percent "cropland" and 2.6 percent "urban," according to official land use statistics,[46] with the remainder as forest, pasture/grassland, parks/designated wildlife preserves and natural habitat. Thus is the national biodiversity cup close to 80 percent full or a little over 20 percent empty? Furthermore, cropland is almost invariably interspersed with wetlands, rough terrain, hedgerows, and other diverse land cover between fields.

To illustrate this point, I will relate my experience when for business reasons I recently drove from my home in Ottawa to both Boston and Washington DC on separate occasions. Within a couple miles of my house, I was on the Canadian equivalent of the US interstate highway system; I headed in different directions for each of the journeys. In total for the round trips, I traveled on highways that took me through parts of two capital regions, two Canadian provinces, and six US states for close to two thousand miles. With the concept of biodiversity in mind, I made a special point of observing the landscape and noted that even in farming districts, there was a significant proportion of land that was either native pasture or, more commonly, forest or other undeveloped countryside. Then there were long distances with little or any serious farming activity. Even in urban centers, I observed biodiversity once I started to look for it. I therefore invite readers when traveling by car, bus, air, or rail to observe the actual share of the land devoted to intensive agriculture with monoculture crops. Even in concentrated corn and grain-growing areas, a surprising amount of biodiversity can be observed.

Although farmers are often accused (falsely, in my opinion) of being environmentally challenged individuals, few casual observers realize that collectively they have committed 34 million acres from 424,000 farms to the Conservation Reserve Program,[47] which is described as follows on the USDA web site for this program:

> *The Conservation Reserve Program (CRP) is a voluntary program for agricultural landowners. Through CRP, you can receive annual rental payments and cost-share assistance to establish long-term, resource conserving covers on eligible farmland.*
>
> *The Commodity Credit Corporation (CCC) makes annual rental payments based on the agriculture rental value of the land, and it provides cost-share assistance for up to 50 percent of the participant's costs in establishing approved conservation practices. Participants enroll in CRP contracts for 10 to 15 years.*
>
> ***Benefits***
>
> *CRP protects millions of acres of American topsoil from erosion and is designed to safeguard the Nation's natural resources. By reducing water runoff and sedimentation, CRP protects groundwater and helps improve the condition of lakes, rivers, ponds, and streams. Acreage enrolled in the CRP is planted to resource-conserving vegetative covers, making the program a major contributor to increased wildlife populations in many parts of the country.*
>
> *The Conservation Reserve Program reduces soil erosion, protects the Nation's ability to produce food and fiber, reduces sedimentation in streams and lakes, improves water quality, establishes wildlife habitat, and enhances forest and wetland resources. It encourages farmers to convert highly erodible cropland or other environmentally sensitive acreage to vegetative cover, such as tame or native grasses, wildlife plantings, trees, filter strips, or riparian buffers. Farmers receive an annual rental payment for the term of the multi-year contract. Cost sharing is provided to establish the vegetative cover practices.*

With the possible exception of the forestry industry and the example in the following paragraph, there is probably no other conservation program anywhere involving the private sector or non-governmental agencies that comes close to matching the scale and impact of the CRP partnership between the USDA and farmers throughout the country. Interestingly, books, articles, and documentaries that dwell on the negative side of food production and seem to care so much about biodiversity and the environment conveniently omit mention of this program of good stewardship of the land covering 31.3 million acres, or 50,000 square miles[48] in all states – an area over one-third the size of Germany.

Another often overlooked but major conservation effort that lends itself beautifully to biodiversity is the work done by Ducks Unlimited[49] in Canada, the United States, and Mexico. This organization has over twelve million acres, or about twenty thousand square miles of wetlands, grassland, and forest under their direct control and management; this is an area nearly twice the size of Belgium. Furthermore, by working with farmers, forest owners, and other landowners, they gave another eighty thousand square miles under a variety of conservation initiatives. Thus in total, Ducks Unlimited is responsible for a natural habitat area on this continent that is half the size of France.

Some critics discount this effort because this conservation organization reaps part of the benefit of their effort by the harvest of some of the waterfowl through hunting. This seems a small price to pay for the maintenance of the large duck and goose population that is not hunted—and even more important, the habitat that is provided for shore birds, fish, frogs, turtles, snakes, insects, small animals, and many other members of the animal kingdom, plus untold species of plants. Furthermore, the wetlands that are preserved or created effectively enhance both flood mitigation and aquifer replenishment. To add to this impressive list, undisturbed wetlands and the surrounding habitat on higher ground play an impressive role in carbon sequestration. Given the area involved in so many locations spread out over the three countries, the overall positive environmental impact throughout North America is most impressive.

After administration and fundraising, 81 percent[50] of their budget goes directly to habitat development. This should be a model for other environmental non-government organizations that all too seldom use funds they raise for actual habitat development or other activities that have a direct positive impact on the environment.

In summary, although farming is a very major, environmentally challenged activity, food production on a large scale is an essential activity that cannot be avoided. However, the rural landscape may be surprisingly more environmentally robust and with a diverse habitat for plants and animals than some writers and commentators would lead us to believe.

9.

Genetically Modified Food: The Controversy

The debate about the health, environmental impact, and ethical issues surrounding genetically modified foods, provides a useful exercise to weigh the pros and cons of this technology and its application. .

Incidentally, should readers be more familiar with the term genetically modified organisms (GMOs)[51], this is the technology that I am focusing on in this chapter. Elsewhere in the book I will use both of the terms to capture the same technology and its application. The concept involves an organism such as a plant, animal, or microorganism that has had its DNA (genetic code) altered by adding foreign genes with the objective to improve performance. Though there are applications beyond food, I will confine my comments primarily to how this science impacts agriculture, with an emphasis on field crops such as corn and soybeans.

A thoughtful and a balanced concern about the potential negative impacts of GM foods are reasonable and appropriate. There are some applications that provoke a negative reaction in most people, such as the project where scientists have apparently spliced the genes of jellyfish into pigs that now glow in the dark.[52] Another is inserting spider genes in goats so that their milk can be spun into super strong fibers.[53] Then there is the herd of two hundred milk cows in China that have been implanted with human genes, which enables them to produce milk that is a near perfect substitute mother's milk for infants.[54] Perhaps there is solid science involved in these applications, but I certainly understand why a great many people tend to be squeamish regarding such technology.

Such unease manifests itself in a general concern that there is something unnatural about GM technology. This is absolutely true, but humans live in a world of unnatural ways in which we transport ourselves beyond walking, communicate electronically at great distances at the speed of light, or the implant of medical devices in the human body. The basic science behind such technologies is quite easy to grasp, and they are adopted with little concern and soon become a "natural" part of our lives.

Unfortunately the rather complex science associated with GM foods is not easily understood by the general population, and its application is carried out during the growing season on farms far removed from the consuming public. These two factors inhibit transparency, and thus there is a natural tendency for apprehension by those who are concerned about something as important as the food we eat. This apprehension, in turn, translates into an anticipated but unknown risk. The natural human tendency when there is undetermined risk is to assume a worst-case outcome.

If risk can be measured, then it is accepted even if it is substantial and potentially catastrophic. The number of deaths and debilitating injuries that occur on the highways per every one hundred million miles driven is a statistical reality that is lamented, but it does not seem to discourage this form of transport. For most, the benefits of convenient personal transportation are deemed to outweigh the understood risk. To continue illustrating this point, every year there are many deaths and even more injuries from falling down stairs—yet no one is advocating banning stairs as a safety precaution, because society has long since concluded that the benefits of stairs far outweigh the slight risk involved with their use. Even the dubious benefits that are associated with a very high health risk are accepted by smokers. In this context I will provide my thoughts on the risk-benefit trade-off for GM foods.

In spite of four hundred million Mexicans, Americans, and Canadians being exposed to GM foods made from corn, soybeans, and canola for at least a decade and a half, there is no incidence of health issues that can be attributed to such crops. In China and elsewhere, hundreds of millions more are similarly including GM foods as a regular part of their diet over many years with no apparent ill effects.

Thus we have a situation whereby GM foods are presumed deleterious to human health by some outspoken people, until such products can be proven to be safe beyond any doubt. Though it is possible to determine if some food or substance is unsafe, the reverse of proving the total absence of risk is virtually impossible. Take tomatoes, for instance, which belong to the same family as the deadly nightshade; they were assumed to be poisonous during colonial times and were only occasionally grown as an ornamental plant. The complete safety of tomatoes as a food was never proven as a scientific reality, but as more and more of the population consumed this product without incident, it became conventional wisdom that tomatoes were safe to eat.

To make matters even more difficult for GM foods, when it comes to balancing risk and benefits, the average consumer is quite likely to conclude that such products are of little advantage in comparison to the equally appealing, non-modified counterpart. Thus it is quite understandable that for many, GM technology is undesirable given an unknown risk without obvious benefits for the individual.

But there are major benefits to GM foods of which the average consumer might not be aware. For starters, GM crops are higher yielding and require less inputs; for North American consumers, this translates into lower food prices than if this technology did not exist. Then there is the need to feed the world, particularly with the combined issues of a growing global population and the uncertainties of climate change. Even the imagined risks of GM food–induced health issues cannot compare to the slow and horrible deaths through starvation or the debilitating and long-term effects of chronic malnutrition.

On one hand there is the no demonstrated and scientifically supported potential risk associated with GM technology, particularly given the trillions of meals already consumed over many years. On the other hand there is the very real benefit of providing adequate nourishment to the global population through the increased yields that are the result of the widespread application of this technology.

There are several balanced papers that list the pros and cons of genetically modified foods. In reviewing these, I remind the reader to note that the advantages are factors such as improved yields, reduced use of chemicals, lower tillage rates, drought and heat tolerance, and other such measurable attributes. The cons, on the other hand, are largely theoretical and subjective. The following is one of the more balanced overviews of both sides of the discussion and is presented on the Human Genome Project Information website.[55] It should be noted that the benefits are mostly proven by demonstrated practices, whereas controversies are often potential concerns that have yet to be proven.

GM Products: Benefits and Controversies

Benefits: Crops

- ***Enhanced taste and quality***
- ***Reduced maturation time***
- ***Increased nutrients, yields, and stress tolerance***
- ***Improved resistance to disease, pests, and herbicides***
- ***New products and growing techniques***
- ***Increased resistance, productivity, hardiness, and feed efficiency***
- ***Better yields of meat, eggs, and milk***
- ***Improved animal health and diagnostic methods***

- ***Environment "Friendly" bioherbicides and bioinsecticides***
- ***Conservation of soil, water, and energy***
- ***Bioprocessing for forestry products***
- ***Better natural waste management***
- ***More efficient processing***
- ***Society***
- ***Increased food security for growing populations***

Controversies

Safety

- ***Potential human health impacts, including allergens, transfer of antibiotic resistance markers, unknown effects***
- ***Potential environmental impacts, including: unintended transfer of transgenes through cross-pollination, unknown effects on other organisms (e.g., soil microbes), and loss of flora and fauna biodiversity***
- ***Access and Intellectual Property***
- ***Domination of world food production by a few companies***
- ***Increasing dependence on industrialized nations by developing countries***
- ***Biopiracy, or foreign exploitation of natural resources***

Ethics

- ***Violation of natural organisms' intrinsic values***
- ***Tampering with nature by mixing genes among species***
- ***Objections to consuming animal genes in plants and vice versa***
- ***Stress for animal***

Labeling

- ***Not mandatory in some countries (e.g., United States)***
- ***Mixing GM crops with non-GM products confounds labeling attempts***

Society

- ***New advances may be skewed to interests of rich countries***

Moving from the concept of risk and benefit, it is interesting to note that virtually all the plants and animals farmed for food and fiber throughout the world have altered genes that have been influenced by humans. All of our grains, fruit, vegetables, poultry, and livestock have been transformed from their wild state to better serve Homo sapiens.

Take dogs for instance, which incidentally qualifies as food animal in some cultures. There is little resemblance between most dog breeds and their wild, ancient relatives, which seem to be tamed gray wolves. The process started twelve to thirteen thousand years ago, probably in the Middle East,[56] and look at what exists today. To choose an example, does a dachshund, which was bred in Germany to hunt badgers, resemble a wolf in size, shape, temperament, and even the fur color and texture? Beyond this one example, there are untold dozens of breeds of dogs that differ as much from one to another as they are distinct from their common ancestor. And yet as different as these breeds are, it is popular to cross these to achieve yet another array of interesting pets.

Human comfort with a great variety of dogs has been around for thousands of years, and yet with such a highly mobile and widespread animal, if there was ever a serious risk of some permanently returning to a wild state, the opportunities to do so have been infinite. However, this has never happened—even packs of feral dogs, as dangerous as they may be, can relatively easily be controlled and are not a significant threat to society as a whole. To borrow the term "Frankenfoods" from the anti-GMO movement that has concerns regarding plants or animals with altered genes escaping into the wild with severe negative environmental consequences, there has never been an emergence of a "Frankenfox" or domestic type of dog causing any serious havoc as a feral animal.

Staying with domesticated animals, food animals such as poultry, cattle, sheep, swine, chickens, turkeys, ducks, and geese are all genetically different, and sometimes extremely so, from their wild ancestors. A mule is the sterile offspring of a male donkey and a female horse, which is an example of a cross-species beast. Few of these domesticated species could live for any length of time or procreate successfully in the wild, and there appears to be no serious environmental issues with any of these despoiling the countryside or messing up the genes of their wild ancestors. One notable exception is the horse, which roams wild in certain parts of North America but to the discomfort of very few.

Another centuries-old practice involving mixing the genetic makeup of plants for food purposes is grafting the roots of one related species to the stock of another, as is routinely practiced in vineyards and orchards. For example, in the nineteenth century the North American plant louse phylloxera[57] was accidentally introduced into Europe and practically destroyed the species of grapes growing there at the time. To this day the only way that European grapevines can resist this bug is to utilize the native North American root stock and graft the vines of the desired grapes to this very foreign species.

Thus one set of genes promoting healthy roots combines with another that delivers the desired grape.

For apples and other tree fruits of this nature the seeds from a McIntosh, red delicious or other apple variety, when planted, do not result in a tree that yields the same desired fruit upon maturity. The only way to ensure that the desired apples will grow is to graft[58] the branches that are known to yield such fruit to the root and main stem of another apple variety that is developed with concern for robustness and not fruit-bearing capabilities. In my own garden, before the maples trees' shade overwhelmed it, I had an apple tree that yielded a half dozen different common apple varieties from branches that originated on the same number of mature parent trees. This procedure is regularly applied to innumerable fruit and ornamental trees.

Though the technology applied here is readily visible to the naked eye and is conducted with regular cutting tools, it is in many ways a variation of the same theme as genetic modification, by splicing genes carried out under a microscope. In both cases such issues as pest control, improved food products, and superior harvest are the result. However, grafting has been applied for centuries and is a practice mentioned in the Bible;[59] it has no detractors today. As time passes the same attitude by the consuming public will probably apply to foods benefiting from genetic modification.

Domesticated food plants such as wheat, corn, soybeans, rice, tomatoes, and potatoes all have been long since altered significantly from their wild counterparts, and they tend to be quite content to remain in the fields in which they are planted and cause no disruption to the natural environment.

To be certain, the process of natural selection based on the random mutation of the genes is technically quite different than those in the modern genetic modification procedure. In the former, people had to wait for the mutation to take place and, through diligent observation, isolate the desired new strain and then propagate it to the point where the population of seeds or animals was large enough to enter commercial agriculture. If there is such complacency about so many existing plants and animals that are genetically different than their counterparts in nature, why draw the line so adamantly regarding the genetically modified version?

The difference between the two processes is that modern geneticists, relying on splicing genes to achieve the desired modification, have taken the guesswork out of the process that previously relied primarily on chance to develop new plants and animals to better feed the world.

With genome mapping advances, scientists have for a number of years been able to conduct a form of natural selection that actually closely mirrors the painstaking steps of early plant breeders, who carefully observed fields of a certain crop to select those plants with the desired superior trait. With advanced molecular marker technology,[60] scientists can determine the desired type of offspring by looking for genes that signal that a certain characteristic, or perhaps group of characteristics, is present. By removing an insignificant part of the outer layer of the seed without destroying it, an instantaneous computer analysis determines if the superior attribute exists with such seeds being segregated for normal propagation.

The advantage of this process over actually observing growing plants is that virtually millions of seeds can be scanned in this manner, greatly improving the odds that the desired traits will be identified and propagated. For those concerned about splicing genes, this process is as natural as the methods used by earlier humans in their quest for more productive crops, animals, and trees. Today the application has spread to all of these areas of breeding.[61] Indeed, perhaps the genetic modification of crops that has become so widespread in the past two decades will soon be eclipsed by advanced molecular marker technology. As an indication of the purity of this approach to plant breeding, the USDA does not require the rather complex licensing procedure now associated with new GM crops.

A few years ago I visited Monsanto's since abandoned wheat breeding facility in the United Kingdom, and I observed this process firsthand. There were several machines about the size of a small desk, each analyzing what seemed like at least a seed a second and selecting the very few with the desired marker for propagation. As a trained agriculturalist who studied in the days of traditional natural selection of living plants, the advantages of this technology as a means to quickly enhance the productivity of wheat and other plants is quite amazing. I predict that this technology will someday be heralded as one of the most important in terms of benefiting mankind during the early twenty-first century.

Returning to the GMO controversy, which also frequently includes the previously described technology, we need a pragmatic, balanced approach to the concept. As an example of such careful reasoning, in 2002 the Royal Society of London[62] (with a membership of 1,400 leading UK scientists) took such a stand on GM foods. Subsequently in 2009 the Royal Society renewed their positive position on GM foods.[63] Furthermore, in 2006 the British Medical Association (BMA) issued the following statement, which captures the sentiment of both institutions:

> *The BMA shares the views of the Royal Society that there is no robust evidence to prove that GM foods are unsafe… However, we endorse the call for further research and surveillance to provide convincing evidence of safety and benefit.*[64]

The UK chief scientist, Professor John Beddington, has also recently provided his full support for GM foods.

> *We need a greener revolution, improving production and efficiency through the food chain within environmental and other constraints. Techniques and technologies from many disciplines, ranging from biotechnology and engineering to newer fields such as nanotechnology, will be needed.*[65]

Also, the UK Society of Toxicology included the following as the first two paragraphs of an executive summary on a paper covering the safety of GM foods.

> *The Society of Toxicology (SOT) is committed to protecting and enhancing human, animal, and environmental health through the sound application of the fundamental principles of the science of toxicology. It is with this goal in mind that the SOT defines here its current consensus position on the safety of foods produced through biotechnology (genetic engineering). These products are commonly termed genetically modified foods, but this is misleading, since conventional methods of microbial, crop, and animal improvement also produce genetic modifications and these are not addressed here.*
>
> *The available scientific evidence indicates that the potential adverse health effects arising from biotechnology-derived foods are not different in nature from those created by conventional breeding practices for plant, animal, or microbial enhancement, and are already familiar to toxicologists. It is therefore important to recognize that the food product itself, rather than the process through which it is made, should be the focus of attention in assessing safety.*[66]

I have chosen these British examples of support for GM foods because the United Kingdom raises only modest amounts of the main GM crops of corn, soybeans, canola, and cotton, and it is not a global leader in the development of this technology. Furthermore the institutions involved are populated by a large proportion of the country's best minds on the subject, dedicated to the well-being of the citizens of the UK. Thus, perhaps unlike the United States, there was little possibility for an industry-influenced second agenda, and the interests were almost certainly only the concerns of public health.

Ongoing research on this issue will continue, but with each passing year the likelihood that any serious health or other risk evolves will recede. To be certain, 100 percent proof will never be achieved, but given the benefits of GM foods in regard to increased yields and reduced inputs, the slight risk element is one that reputable scientific and medical entities seem prepared to accept.

Yes, some environmentalists and others are trying hard to discredit GM technology, but peer-reviewed articles supporting this in the respectable medical, scientific, or agricultural literature are rare. For example, there is an experiment that is often quoted by anti GMO groups that revealed monarch butterflies that were fed pollen from genetically modified corn had high mortality rates. When peer reviewed by other scientists, it became apparent that these migratory insects lived on ragweed pollen, and the corn variety was not on their menu. An analogy would be to offer a carnivore with meat as the normal diet genetically modified corn only and, when it did poorly or perhaps perished through malnutrition, declare the experiment proved that the grain posed a serious health risk that could be extrapolated to humans.

There are some who have made a career out of taking on the role of expert who are against biotechnology and its application in agriculture in spite of no apparent formal training or previous professional experience in this area. Such one-sided fear mongering is not at all helpful to fostering reasonable discussion on the matter. In response to such pundits, an organization called Academics Review with the tagline "Testing Popular Claims Against Peer-Reviewed Science" has painstakingly taken most of the many claims about GM hazards and placed them side by side with a host of existing peer-reviewed research and papers prepared for other purposes but that thoroughly debunk most claims with solid science.[67]

I can fully appreciate the business model of individuals and environmental, non-government organizations that are against GM foods. Though they may sincerely believe that the practice is badly flawed, it is a fact that their ability to raise profile for themselves or their organization, or more importantly raise funds through donations from the public, depends on a dramatic, negative position without any deviation from this cause. It would be a betrayal to their organization if suddenly they had an epiphany and proclaimed support for GM foods. Not only would funding for the causes evaporate, but credibility would suffer in the eyes of those whom they have convinced their crusade is a just one. It is therefore a reality that opposition to GM technology is entrenched for understandable reasons, and it is unlikely that this position will be altered any time soon.

But as Senator Daniel Patrick Moynihan once said, *"Everyone is entitled to his own opinion, but not his own facts.*" I raise this because for an issue as important as GM foods, there are moral issues regarding misrepresenting the facts and causing unnecessary concern among the casually informed but rightfully concerned public. This concern also manifests itself in delaying progress in developing this essential technology to feed the world. One example of this was the unfortunate delay in developing GM wheat by perhaps two decades.

Although the acreages in GM crops are huge, the number of food crops involved is surprisingly small, with corn, soy beans, canola, cotton (with cottonseed oil as part of food), rice, sugar cane, and sugar beets representing the important crops that are enjoying the benefits of this technology. The major traits involved are also few in number, with pesticide tolerance, insect defense mechanisms, enhanced nutritional content, drought tolerance and efficient water uptake, and fertilizer use efficiency as the most common beneficial attributes. Of the type of foreign genes involved, some are quite interesting.

For example, daffodil genetic material was used with conventional rice to enhance the vitamin A content and improve the diet of millions. The following quote from an article prepared by the European Food Information Council presents a clear indication of the value of this one particular GM trait in a staple food.

> *Over 100 million pre-school age children suffer from vitamin A deficiency, as do millions of women of child-bearing age. Vitamin A is essential for the operation of the body's immune system and is responsible for protecting mucous membrane cells. Vitamin A deficiency causes increased risk of infection, night-blindness and, in severe cases, total blindness. Over 1 million children die every year as a result of vitamin A deficiency.*[68]

I find it difficult to understand those who have zero tolerance for GM foods in light of such convincing benefits for those most in need of the nutritional advantages that this technology brings.

Another important benefit of GM crops is the weed killer tolerance trait, which is particularly important because it enables farmers to utilize herbicides to control weeds, leaving robust food plants. However, before I describe the yield, energy savings, and environmental advantages of such herbicides, a few paragraphs are probably necessary on the health controversy around this group of agricultural chemical, which includes Roundup. As can be expected, there are two points of view, both for and against, on health grounds (with birth defects and incidence of cancer as two of the most frequently referred to issues). Without getting into the complexities of such studies, I shall take a different approach and draw upon highly reputable statistical evidence on actual cancer and birth defects in various locations in the United States.

I refer to the National Cancer Institute[69] state profile for Iowa, which arguably has among the highest exposure of its citizens anywhere to Roundup and similar weed killers, given the very high concentration of corn and soybean production. The state incidence of cancer is 472.3 per 100,000 inhabitants annually for the years 2003–2007; the US national average is a nearly identical 464.5. Furthermore, the highest incidence in the state is Page County, with an occurrence of 541.2, and the lowest is Ringgold County at 394.6. Utilizing data from the city-data.com website, Page County is 65 percent urban and 35 percent rural, while Ringgold County is entirely rural. Given that farmers in a rural setting will have a greater exposure to Roundup than their urban counterparts, these statistics do not seem to support the thesis that there is a causal effect between contact with this weed killer and the incidence of cancer. To further illustrate this point, New York State has a near identical annual cancer incidence rate of 483.2 per 100,000. With the majority of its citizens having no exposure to farms or weed killers, this again seems to indicate that cancer and Roundup have little to do with each other.

The other main health issue related to such herbicides is the concern that it causes birth defects. Here again population statistics seem to tell a somewhat different story. According to the National Birth Defects Prevention Network, Iowa has an incidence of 75.13 birth defects per 10,000 live births for the years 2002–2006, whereas the US average is somewhat lower at 60.70. While this is close to fourteen more unfortunate births per ten thousand, epidemiologists probably would not consider this to be an indication of some overwhelming risk to the population. Furthermore, when one looks at the low birth defect rates in such states as Arizona, New Mexico, Maine, Vermont, and others with modest industrial (as well as agricultural) activity compared to other regions in the country, Iowa statistics are more in line with states with higher levels of chemical exposure of all types.

Now, what does the ability to kill weeds competing with food crops really mean back at the farm? Environmentally, the big plus is that excessive tillage to control weeds is largely a thing of the past. The near elimination of this practice reduces wind and water erosion and maintains a more natural subsurface profile for the living organisms that thrive there. Reduced tillage also conserves water because there is plant residue left on the surface that reduces evaporation, and also undisturbed soil does not suffer from the burst of evaporation that occurs when tillage takes place.

Less tillage also reduces fuel consumption, and thus the carbon footprint of crop production drops dramatically. The cost of production goes down, and yields are enhanced. Also the use of Roundup and similar chemicals provided by Monsanto and their competitors reduces or eliminates the application of other herbicides that have been utilized for decades.

To be sure there is substantial uneasiness, primarily in non-farming circles, regarding Monsanto's exceptional brand recognition which makes the company the "poster boy" that some none farm folks love to hate. What many do not realize is that Monsanto has many of competitors such as Syngenta, BASF, Du Pont, and Bayer that offer many of the same products and services. Take Syngenta for instance. According to their web site they offer 24 different herbicides (including a perfect substitute for Monsanto's "Roundup" called "Touchdown"), 37fungacides, 24 insecticides and 27 seed treatment chemicals plus a large selection of corn, soybean and other seeds, some of which are genetically modified. Furthermore they have 26 thousand employees in 90 countries. It is thus a complete urban myth that farmers are in any way tied to Monsanto for their Roundup, any of their advanced seed varieties or other agricultural inputs.

To use an analogy, if a motorist purchases or leases a new GM automobile they are locked into the company for their motoring needs for at least three years. This is much less freedom of choice than Monsanto clients have who are only locked into any arrangements for the growing season. Indeed it is quite possible to mix and match in the farming industry and use two or more competitors' seeds and chemicals on the same farm.

For several years I was a member of the Technical Committee of the US National Association of Wheat Growers,[70] a group of about two dozen representatives of industry involved with technology supporting wheat production. The Monsanto member of this committee was always a welcome contributor by as were other agricultural chemical and seed providers listed above. There was a fundamental interest on the part of farmer members of this group to learn of the latest developments. They have a reason for doing so—adopting the latest technologies makes them money, which is totally contrary to the concept many urban folks have regarding Monsanto.

It should be pointed out that farmers everywhere have a ready option to grow non-GM crops, so those who do choose to go this route do so willingly given that the technology works well for them and enhances their profits. In this regard the concern that Monsanto and others' technology is not helpful to small-scale farming, particularly in developing countries, is quite appropriate but possibly misdirected. Companies everywhere have their market niche. For example, Toyota is not criticized for having little interest in bicycles or other basic transportation technology that would be appropriate for those who cannot afford a car. So too is it with Monsanto: their market is primarily associated with larger scale agriculture, and thus they should not be considered to be all things to all farmers.

However, this fundamental fact does not alleviate the concern that small-scale farmers, particularly in the tropics, are without a champion regarding advanced agricultural technology. Historically such developments have been fostered by foreign assistance, but this has dropped off dramatically primarily because of complacency resulting from the spectacular yield gains in a host of countries, which were achieved through the Green Revolution from World War II until the 1970s. To illustrate this point, official development assistance or foreign aid to developing countries for agriculture research dropped by 61 percent between 1980 and 2003[71]; the trend was particularly pronounced in rural Africa. Instead of lamenting failure by the private sector, focus on the issue might be better directed to reversing the trend and quickly increasing foreign aid for appropriate agriculture technology development in the Third World. After all, the role of foreign aid is to assist developing countries where the private sector has little interest. Regarding agriculture assistance, aid agencies have failed badly and have done so for an entire generation.

An important theme of this book links the need to feed the world as a fundamental responsibility of agriculture. Here is where GM foods make a big difference and will continue to do so. According to USDA statistics,[72] the average national corn yields increases from 134 bushels per acre in 1999 to 154 in 2009. During the same period, wheat yields increased a modest 43 bushels per acre to 45. Corn productivity increased by a factor of 20 bushels per acre, compared with only 2 bushels for wheat, as the later enjoys none of the yield benefits provided by GM technology. In percentage terms, corn yields increased by 15 percent, and wheat lagged with only a 5 percent improvement.

What is even more dramatic is the thirty-year yield improvement that saw the amount of corn that was harvested increase from around 90 bushels per acre in 1980 to over 150 by 2010. [73] Much of this increase can be attributed to GM technology. I can think of no other time in human history where a major crop has shown a consistent, near 2 percent growth a year, and there is a strong likelihood that this trend will continue with even more improvements, thanks to advanced plant breeding.

Regarding total production, wheat has remained nearly constant during this period, whereas total corn production has increased by a third according to the USDA. Globally the relationship between the two is very similar, with wheat increasingly left behind. Given that wheat is used almost exclusively as a human food and corn is used primarily for livestock feed and energy, this does not auger well for a growing global population. Had GM wheat been available, the yield trends would have likely been similar.

Indeed it was because of this decline in the fortunes of the wheat industry, along with the shift to the more profitable genetically advantaged corn and soybeans, that the farmer-directed National Association of Wheat growers and their Australian and Canadian[74] counterparts recently encouraged Monsanto, and their competitors to concentrate on developing GM wheat varieties. Though they realize that there will be some international consumer resistance, they have determined that the increased yield and therefore exported food will more than make up for any market pushback. Also, given the major role in the global marketplace, the three countries' wheat industries, working in unison, will still remain dominant suppliers given that many major markets do not have a significant aversion to GM foods, because corn has already met widespread consumer ambivalence, if not outright acceptance.

Also in support of GM crops and the improved yields is the reduction in the need to bring the last remaining virgin land under cultivation to meet ever growing world population, which is demanding more meat and animal proteins. In its own way, GMO technology is an unheralded but effective means to save the trees in tropical rainforests and elsewhere.

The debate about the safety of GM foods will continue unabated, and for the wealthy in Europe, North America, and Japan, efforts will be made (on an individual basis at least) to attempt to avoid GM products in their diets. If they are willing to pay enough, the market will look after these folks who have every right to choice.

There is a continuing call on the part of consumer groups for foods with GM ingredients to have a statement on the product indicating the presence of such ingredients. Absent any evidence that GM foods are harmful, such a notice would incorrectly signal a food safety issue that does not seem to exist. Indeed, given all the controversy regarding GMOs, some of which is incorrect but very negative, many consumers would have unfounded concerns because most packaged foods would be difficult to certify as absolutely GM free. There seems to be a lot more pressing and clearly known food concerns that improved labeling might address before the GM issue should be highlighted. Indeed, certified organic foods, if genuine, are GM free, so consumers do have a choice if they wish to exercise it.

In summary, regarding the overarching importance of feeding the world, it should not be overlooked that to a hungry person, food from GM crops is immensely better than no food at all. Given this fundamental fact and the existing widespread consumption of genetically modified foods by hundreds of millions of humans and animals for nearly two decades without any documented health problems, it is highly probable that genetically modified crops are here to stay and will continue contributing to the reduction in hunger.

10.

Organic Foods—More Controversy

There is much that is encouraging about organic foods, such as sustainable farming practices, reduced chemical inputs, opportunities for smaller farms to achieve better prices for their efforts, and perhaps a superior food product. I will not dwell on these positive attributes because there are numerous sources that cover this topic in a very thorough fashion. However, in spite of claims that this approach to farming is a panacea for the avoidance of chemicals in food production, healthy eating, and feeding the world's population, the organic food industry does have some issues of which readers might not be aware.

Though many of the following comments are critical of some elements of this movement, I wish to emphasize that there are legions of small organic growers for which their approach to this type of farming is akin to a religion. For consumers with direct access to such sources of food, the chances are high that the products purchased will fully measure up to their organic expectations. But I maintain that such dedication may not be universal, and thus with some exaggeration, I parody the lament of youth from an earlier era: "Do not trust an organic farmer with over thirty acres!"

Like any other commercial activity, farming is a business where at the end of the day, there should be an acceptable level of profit to compensate for the effort involved. Even with the best of intentions, if during the growing season there is a weed, fungus, or insect infestation that threatens to damage or even destroy the crop—and thus income for the year—is there not a temptation to bend the rules and rely on a pesticide that the farmer across the road successfully utilizes on a regular basis, and to supply food without any consequences regarding food safety? Given that the industry is regulated by certification based on the honesty of the famer and the integrity and capability of a local certification agency established for this purpose, the system is not as foolproof as some consumers might assume. For example, during a troubled growing season, would it not be in the realm of the possible for an organic farmer to apply badly needed chemicals to save not only the crop but also to preserve the well-being of his family?

It may surprise some that the only enforced food safety regulations for this element of the industry are the same as for non-organic foods and are carried out by the Food Safety Inspection Service (FSIS) of the USDA and the Food Inspection Agency in Canada. Neither of these have the resources or the mandate to ensure that organic foods are produced fully in accordance with the established practices required for certification.

The FSIS does not have unique inspection protocols for organic food but rather relies on an organic certificate from an accredited certifying agency of which, according to the website of the USDA National Organic Program (NOP) website,[75] there are fifty-three domestic and forty-one foreign entities.[76] These are local in nature, and one might assume they have a much closer relationship to the grower than the FSIS ever could. This would be particularly the case with the foreign certification bodies, some of which may be ethically challenged.

As there is no established or universal inspection regime at the certification agency level, the potential for bending the rules in the fields or plantations certainly exists. When I encounter a manufactured food product that claims to be organic but was obviously produced in volume, involving many ingredients from different sources, the potential for non-organic products slipping into the mix is far from remote. For example, even such basics as organic bananas and other tropical fruits perhaps originate in jurisdictions with less than perfect records for regulatory enforcement of any kind, and this raises even further doubts regarding strict adherence to organic production methods.

Although the National Organic Program does have responsibility for suspending organic food producers and also certification agencies, revoking their status both domestically and abroad, their record for doing so does not seem in balance with the number of farmers and agencies involved. For example, during the first eleven months of 2010, the NOP suspended 224 organic producers, domestic and foreign. Of these only 21[77], of which domestic organic farmers accounted for 15 incidences, had infractions such as the inappropriate use of substances or other practices that were not in accordance with organic farming. The remainder had been suspended for administrative reasons unrelated to farming practices, usually because they did not pay their annual fees. Of the nearly thirteen thousand certified organic growers[78] supplying the US food industry this seems like a very small number of actual infractions.

To draw an analogy, assume a community had a population of thirteen thousand and had a police force of fifty-three members (to match the number of certifying agencies). In terms of law and order, less than one policeman out of three would make but a single arrest over an eleven-month period. Should such a community exist, one would conclude that the arm of the law was not very long, or the population was uniquely law abiding and did not need policing.

Although unrelated to organic food certification, there is evidence that mislabeling of food products is relatively common and probably deliberate. According to a newspaper article[79] on the application of DNA testing of food to establish product integrity, it was noted that over a nine-year period the US government determined that 37 percent of all fish and 13 percent of shellfish were mislabeled. While those in the organic food industry may have higher standards than fishmongers, the fact remains that mislabeling to enhance the value of a product is not uncommon in the food industry. Given that providence is extremely difficult to determine for organic foods, consumers should be aware of the possibility for mislabeling, particularly in light of so little evidence that there are vigorous enforcement procedures in place.

It should also be pointed out that Fair Trade Coffee has critics who maintain that the opportunity to mislabel the product to attract a higher price, with little or no benefit to the small remote grower, probably exists in that trade. The rather opaque journey of the coffee bean from rural areas of countries that get less than top marks on Transparency International's corruption index to a coffee shop or supermarket is one that lends itself to product mislabeling. Some retailers, to protect their reputation, will have rigorous oversight protocols in place, but others may be less scrupulous.

Returning to the world of US certification agencies, upon closer inspection there were only ten agencies that actively exposed one or more wayward organic farmers during the first eleven months of 2010. This means that 43 certification agencies, with an average of 250 farmers per agency to oversee for organic compliance, found nothing sufficiently untoward in their practices to disqualify them. Given that all parties are operating in an environment where noncompliance would almost certainly not infringe on any enforceable laws, and that the temptation to do so certainly exists if crop losses are looming, this whole scenario of apparent near perfect behavior in such a population of organic growers does not seem realistic.

Similarly the failure rate of certifying agencies seems to be very modest. For the five-year period between August 2005 and January 2011, only two certification agencies had their status revoked.[80] I conclude from the very limited number of suspension farmers and revoked certification privileges that the NOP does not have the means or the mandate to vigorously screen even the basic players in the organic movement.

Also, in spite of the profile, organic farming is not as widespread as one might think. As seen in the following table, only 1.5 percent of all farms in Canada are fully certified as organic.[81] Given that most organic farms, particularly the ones who are managed by true believers in the cause, tend to be small, the total production of organic foods is substantially less than the proportion of total farms might indicate.

Farms producing organic products, by certification status, Canada, 2006

Certification status	Number of farms reporting	Percentage of all farms in Canada
Organic but not certified	11,937	5.2%
Certified organic	3,555	1.5%
Transitional	640	0.3%

Regarding the United States, the following is from the website of the Economic Research Service (ERS) of the USDA. It also places the scale of the organic movement in the context of all farming operations in the United States.[82]

> *Overall, certified organic cropland and pasture accounted for about 0.6 percent of U.S. total farmland in 2008. Only a small percentage of the top U.S. field crops—corn (0.2 percent), soybeans (0.2 percent), and wheat (0.7 percent)—were grown under certified organic farming systems. On the other hand, organic carrots (13 percent of U.S. carrot acreage), organic lettuce (8 percent), organic apples (5 percent) and other fruit and vegetable crops were more commonly organic grown in 2008. Markets for organic vegetables, fruits, and herbs have been developing for decades in the United States, and fresh produce is still the top-selling organic category in retail sales. Organic livestock was beginning to catch up with produce in 2008, with 2.7 percent of U.S. dairy cows and 1.5 percent of the layer hens managed under certified organic systems.*

The USDA also conducted a comprehensive survey of yields and acreages of a host of organic crops in 2008,[83] which included twenty-three fruits/nuts/berries, nineteen field crops, and twenty-one different vegetable types.

Of the sixty-three organic crops surveyed, fifty-nine had a yields loss over conventional agriculture. Many of the losses were quite significant, with important crops such as oranges having a 60 percent reduction in yields for organic compared to the national average. The three major grain crops also had major yield reductions, with corn down 28 percent, soybeans 34 percent, and winter wheat 40 percent. For vegetables it was similar, with lettuce yields down 70 percent, carrots 64 percent, and tomatoes 35 percent.

To put this into perspective, the total US corn harvest in recent years is slightly above twelve billion bushels, of which about four billion bushels, or one-third, are used to produce ethanol.[84] Thus if the US corn crop was to move to 100 percent organic, the 28 percent reduction in the yield would be close to the amount of this grain that is utilized to produce ethanol.

The economic trade-off in organic farming is yield loss, which is offset by a substantial premium in the prices consumers are willing to pay for food, plus probable reduced input costs. The following table from the Ontario Ministry of Food and Rural Affairs[85] provides a rather precise comparison of organic production yields and revenues, which shows a rather healthy profit advantage per acre for organic growers. While a few organic food production websites report yield improvements for certain crops and growing conditions, there is a near universal consensus that organic farming yields are significantly less than for conventional agriculture.

Organic Crop Returns and Non-Organic Crop Returns

Organic Crop Returns			
Crop	Yield	Price	Estimated Gross Margin
Soybeans	30 bu/ac	$16/bu	$281/ac
Corn	98 bu/ac	$7.25/bu	$375/ac
Winter wheat	60 bu/ac	$8.25/bu	$251/ac
Spelt	1.1 tonne/ac	$400/tonne	$186/ac

Non-Organic Crop Returns			
Crop	Yield	Price	Estimated Gross Margin
Soybeans	40 bu/ac	$8.00/bu	$131/ac
Corn	130 bu/ac	$4.00/bu	$146/ac
Winter wheat	75 bu/ac	$4.50/bu	$162/ac

While it is mere speculation on my part, the numbers on this table do give rise to an opportunity for the less scrupulous to make a lot of money. Take soybeans for instance. If a declared organic farmer were to follow the same perfectly legal practices of his conventionally farming neighbor, his yield would increase by ten bushels per acre and return by $160 providing him with a total of $391 per acre or over three times what the grower across the road would receive. Given the lax oversight of the organic movement would this not be tempting?

One might ask, so what if yields are less, as long as the farmers receive a better price, while at the end of every crop year there is a substantial carryover of conventionally grown grains, and at least all commercial markets are fully supplied with what is required by the food industry. Furthermore, there are some demonstrated advantages for the environment that authentic organic farmers can probably demonstrate. Given the very modest proportion of the global agricultural resources dedicated to organic production, there is no harm whatsoever in providing consumers with what they want.

However, I tend to disagree with those who contend that this form of agriculture can realistically serve a larger purpose as a major player in the overarching task in feeding the world. At the margins of agriculture, this practice fills a niche market, primarily for the prosperous, but if it became a mainstream means of farming, global food security could be challenged. To illustrate this point, I refer once more to the chart above. For every acre of organic wheat produced in Ontario, there is a loss in yield of fifteen bushels, or about enough calories to feed someone for a year. When elsewhere in this book I provide some detail on the global situation of food security, it is clear organic farming is not the solution and may be seriously detrimental if, as proponents advocate, it expands substantially in such major food-producing countries as Canada and the United States.

There are well-meaning individuals and organizations that maintain yields actually are better than conventional agriculture through organic farming methods. They have individual farmers with yield numbers to prove their point, but does the exceptional producer really represent the average organic farm as is covered in the Ontario statistics for the entire movement?

In such instances, high yield claims for organic operations probably do not account fully for the labor input. For example, while my grandmother was not an organic grower, her practices would have been very close when she grew a garden that on a square foot basis clearly produced more food than on the family farm where we practiced conventional agriculture. Every spring my father would provide a load of composted chicken manure that was duly dug into the soil by hand, During the growing season at the slightest hint of dryness out came the garden hose. Weeds never had a chance beyond sprouting and in the autumn blankets covered the more sensitive plants against an early frost. The vegetables were selectively collected at their peak of maturity which meant that for a few weeks there was a small harvest most days. This was clearly a delightful hobby for my grandmother who had the time available for such an undertaking. I relay this childhood observation as claims of high yields by organic growers may be the result of labor intensive activities closer to my grandmother's scenario than one would ever observe in mechanized farming or could be applied if anything approaching a decent wage was expected by those involved.

On the positive side for organic, the USDA national survey of sixty-three crops had four examples of better yields with such farming. These were hay at 10 percent over conventional farming, plums at 3 percent, pecans at 12 percent, and sweet potatoes at 23 percent. For the perennial orchard crops of plums and pecans, weed control is of a much lesser issue than for annual crops (herbicides are not required), and also advanced plant generics is not widely applied so both conventional and organic farms tend to harvest the same trees. Because of the legumes in the hay crop which is another perennial, nitrogen is captured by the plants themselves from the air and also where there is hay, manure is more frequently available as another natural fertilizer. The point here is that some plants lend themselves to organic farming but considering their share of production as a percentage of the total national output, these four crops are not particularly significant.

Given that prices for the products are definitely better for organic, and that depending how labor costs are factored in, a good argument can be made that inputs are also on the better side of the ledger for such foods. If yields are also better, as some maintain, would not conventional agriculture, which is constantly on the lookout for improved output per acre, embrace such advantageous practices? After all, organic farming is not generally viewed to be a religion, and in any event conventional farmers tend to be an agnostic bunch as far as their trade is concerned and thus will pick up best practices wherever they can find them. To conclude, organic agriculture does not seem to be a viable approach to produce more food off an acre of ground land unless substantially low-priced human labor is involved.

More than once I have mentioned the human labor factor in organic farming as an important resource. In some places in rural China, where direct personal effort and not machines remain very much part of rural farming, the yield in vegetables per acre can be impressive. This labor is available under these circumstances because the opportunity elsewhere for these farm workers remains minimal. In North America, I sincerely admire anyone who wishes to produce food by the sweat of their brow—it can be immensely satisfying, but is it really a vocation to produce substantial quantities of food, or is it rather a somewhat profitable hobby?

Another factor impacted by this loss of yield of organic crops is the concept of indirect land use, which is applied to corn used for ethanol production. The theory works this way: for every acre taken out of food production, another must be found somewhere on the planet to keep up with feeding the global population. It is an imprecise science, but the logic has some merit. Continuing with this concept, as more land transitions into organic food production with demonstrated production losses on a per-acre basis, then in real terms this could mean the loss of Brazilian or Indonesian rainforest or other virgin vegetation elsewhere.

I recently had a conversation with someone who was born in Sub-Sahara Africa and now makes a professional living in Canada. He made the rather interesting statement: "All we have to eat back home is organic foods, and yet we are starving! Farmers do not have advanced seeds, fertilizer, or chemicals, so by default all food produced is organic. The only problem is that yields are very poor." I am fully aware that this scenario has little to do with the reality of organic farming in North America, but there is a universal downside regarding yields with this type of agriculture.

To be certain, organic agriculture is significantly less dependent on chemical inputs than conventional farming. However, it may surprise some that there are ranges of "natural" pesticides and other chemical inputs that are regularly utilized and quite acceptable for certification as a fully compliant organic food product. The following table of approved substances that may be used in organic farming is posted by The European Crop Protection Association (ECPA). Some of these, particularly copper, is cited as being deleterious to the soil.

Approved Substances for Organic Farming in the European Union

Spinosad	x
Pheromones	x
Deltamethrin	x
Lambdacyhalothrin	x
Ferric phosphate	x
Copper	x
Ethylene	x
Quartz sand	x

Sulfur	x
Calcium hydroxide	x
Potassium hydrogen carbonate	x

Source: European Crop Protection Association (ECPA)[86]

For the US, the following is a list of approved sanitizers and cleaners which organic growers can use for certified products.[87]

- *Acetic acid. Allowed as a cleanser or sanitizer. Vinegar used as an ingredient must be from an organic source.*
- *Alcohol, Ethyl. Allowed as a disinfectant. To be used as an ingredient, the alcohol must be from an organic source.*
- *Alcohol, Isopropyl. May be used as a disinfectant under restricted conditions.*
- *Ammonium sanitizers. Quaternary ammonium salts are a general example in this category. Quaternary ammonium may be used on non-food contact surfaces. It may not be used in direct contact with organic foods. Its use is prohibited on food contact surfaces, except for specific equipment where alternative sanitizers significantly increase equipment corrosion. Detergent cleaning and rinsing procedures must follow quaternary ammonium application. Monitoring must show no detectable residue prior to the start of organic processing or packaging (example: fresh cut salads).*
- *Bleach. Calcium hypochlorite, sodium hypochlorite and chlorine dioxide are allowed as a sanitizer for water and food contact surfaces. Product (fresh produce) wash water treated with chlorine compounds as a disinfectant cannot exceed 4ppm residual chlorine measured downstream of product contact.*
- *Detergents. Allowed as equipment cleaners. Also includes surfactants and wetting agents. All products must be evaluated on a case-by-case basis.*
- *Hydrogen peroxide. Allowed as a water and surface disinfectant.*
- *Ozone. Considered GRAS (Generally Regarded As Safe) for produce and equipment disinfection. Exposure limits for worker safety apply.*
- *Peroxyacetic acid. Water and fruit and vegetable surface disinfectant.*
- *Carbon dioxide. Permitted for postharvest use in modified and controlled atmosphere storage and packaging. For crops that tolerate treatment with elevated* CO_2 *(≥?15%), suppression of decay and control of insect pests can be achieved.*

- *Wax. Must not contain any prohibited synthetic substances. Acceptable sources include carnuba or wood resin waxes.*
- *Ethylene. Allowed for postharvest ripening of tropical fruit and degreening of citrus.*

As ethylene is an accepted substance on both lists, I will delve into its application and source in some detail. It is a widely used industrial chemical that is applied to fruit to hasten ripening by destroying the green chlorophyll cells to force the ripened color, such as a perfectly colored orange, which appeals to the consumer. Though it is a natural, gaseous product slowly released by maturing plants (and impractical to capture), the American acceptance of this substance does not specify the source as organic as it does for some other compounds. Thus the ethylene used for ripening purposes comes from a variety of industrial process as described by ICIS, the British chemical industry monitoring organization, which is described as follows.[88]

> *"Ethylene is produced commercially by the steam cracking of a wide range of hydrocarbon feedstocks. In Europe and Asia, ethylene is obtained mainly from cracking naphtha, gasoil and condensates with the coproduction of propylene, C4 olefins and aromatics (pyrolysis gasoline). The cracking of ethane and propane, primarily carried out in the US, Canada and the Middle East, has the advantage that it only produces ethylene and propylene, making the plants cheaper to construct and less complicated to operate."*

Though there is nothing that would be deleterious to human health by food being exposed to ethylene, it does strike me that to artificially ripen organically produced food with a gas made from a petroleum feedstock is somehow not in the spirit of the organic movement that consumers have come to understand. Also, using ethylene produced from propane for cosmetic purposes and yet reject nitrogen fertilizer manufactured from natural gas which would enhance yields and feed more seems to reflect convoluted logic. And while on the topic, bleach as an approved substance may also surprise some who are of the opinion that organic means chemical-free food.

Incidentally for manufactured organic food products, sodium chloride or common salt is not regulated in the prepared products. Those concerned about their salt intake for health reasons should be aware of this issue as "organic" only addresses the production of the basic farmed ingredients and not non food items that are added during processing.

Also, pyrethrins[89], which is a natural insecticide derived from a variety of chrysanthemum, is an approved substance for organic farming in the above list, but it is far from being a benign substance for humans and other animals, such as birds and of course beneficial insects. A study which appeared in the *Journal of Pesticide Reform* included the following quote.[90]

> *"According to a U.S. Environmental Protection Agency (EPA) survey of poison control centers, they (pyrethrins) cause more insecticide poisoning incidents than any other class of insecticides except the organophosphates."*

The second and recent study conducted over two years by the University of Guelph, one of Canada's leading agriculture degree–granting universities in collaboration with the University of Kansas, concluded that the selective synthetic pesticides were less destructive to a wider range of insects, including the beneficial ladybeetle, than their natural counterparts.[91] The logic behind these results is quite basic: synthetic pesticides are designed to be selectively effective by perhaps impacting on the nervous system of only targeted insects, but they will prove non harmful to some or perhaps many of the benign or beneficial species. On the other hand, pyrethrins and mineral oils (also allowed in organic farming) kill virtually all insects when exposed to these materials.

Besides insects, pyrethrins are also very toxic to fish, and because the substance can adhere to fresh fruit and vegetables, the US Agency for Toxic Substances & Disease Registry states it has the following effect on humans.

> *"Pyrethrins and pyrethroids are insecticides that are applied to crops, garden plants, pets, and also directly to humans. High levels of pyrethrins or pyrethroids can cause dizziness, headache, nausea, muscle twitching, reduced energy, changes in awareness, convulsions and loss of consciousness."*

In spite of my comments on the questionable use of pesticides and issues surrounding the reliability of the certification procedure, I consider many of the practices of organic farming to be a positive contribution to agriculture. However, what I do find challenging to accept is the unambiguous position that some adherents of the movement maintain that organic means near perfection on the path from field to fork while conventional agriculture is quite the opposite, with food from this source being something to be avoided.

This leads me to the point that might be more practical: perhaps it may be worth contemplating that "organic" should not be certified as relying not on only natural inputs without any exceptions, but rather using a "best practices" approach. In such a regime, organic pesticides would be tested in an objective scientific manner alongside the synthetic counterparts to determine both the safety aspects of the resulting food products and also the impact on desirable insects and other wildlife. One could draw a parallel to the field of medicine, where there are both natural and synthetic remedies that are generally judged on their efficacy and safety based on peer reviewed science.

The proponents of organic products quite naturally work diligently to embellish their brand. In this context it is perfectly understandable for them to place the worst possible stigma on conventional agriculture to enhance brand superiority for their range of food products. I therefore invite readers to do a bit of research on the bona fides of those who are against conventional agriculture to determine if there is something in their background that might mean some association with the organic movement and therefore a bias and lack of objectivity.

Indeed the assumed nutritional benefits of organic foods over their conventional counterparts may be an example of this brand embellishment. In the September 2009 edition of the *American Journal of Clinical Nutrition,*[92] the rather innocuous sounding paper "Nutritional Quality of Organic Foods: a Systematic Review" was published. The study was conducted by researchers at the UK-based London School of Hygiene and Tropical Medicine of the University of London, and it involved a systematic review of 162 research papers published from 1958 to 2008 on the comparison of the nutritional aspects of organic versus non-organic foods. Of these papers, 55 were of a satisfactory quality to be included in the final review. The overall and seemingly objective conclusion was that from a nutritional point of view, organic foods were not superior to conventionally grown or raised products.

I appreciate that for the champions of organic foods, this is akin to saying "God does not exist" to a believer. Though other studies exist that may demonstrate a more favorable outcome for the organic movement, this overview, given the source and scope, is difficult to dismiss, particularly as the researchers declared that "The funding organization had no role in the study design, data, collection, analysis, interpretation, or writing of the report." Furthermore, it was supported by the UK Foods Standards Agency,[93] which is in their words, *"an independent government department set up by an act of Parliament to protect the public's health and consumer interest in relation to food."* As a tag line they include, *"Everything we do reflects our vision of safe food and healthy eating for all."* It is very doubtful if either organization had an agenda other than to establish the best objective overview of organic food nutrition in comparison to conventionally produced farm products.

For further reading on the more controversial side of organic food production and consumption, I recommend the chapter about this issue from the book *Just Food* by James McWilliams.[94]

I would like to conclude this chapter on a positive note for organic foods and salute the countless farmers who practice this type of agriculture so diligently. These folks do the best they can to bring us what consumers ask for in the marketplace, and they are exceptional stewards of the land at the same time. I just question those who think this approach is a panacea for providing what they might perceive as a solution to feeding the word. Also the consideration that non-organic foods are somehow to be avoided adds unnecessary confusion and anxiety for the consuming public.

For those who can afford organic foods and wish to focus on this source of nutrition, it is a lifestyle choice and a perfectly good one. However, for the majority of consumers who enjoy non-organic foods, there should be no undue guilt or concern regarding the quality and safety of the food they feed their families.

Regarding the issue of global food security, it is much better to have non-organic food on the fork than no food at all.

11.

The Hundred-Mile Diet

When I was growing up on the farm, the concept of the hundred-mile diet was unknown, but during the summer months we frequently ate off the land, which was in essence a hundred-yard diet because everything came from the garden and farm. A typical summer meal might be spring lamb, new potatoes, sweet corn, and a host of other vegetable possibilities, with dessert being freshly picked raspberries and whipped cream from our cows. At the time I was quite unimpressed and I thought lamb and raspberries were things that farm kids had to eat because there was nothing better available. My wife, who comes from an East Coast fishing community, had a similar childhood disdain for lobster, which also changed dramatically when she left home.

My quest for local has been reignited and goes on to this day. During the summer months, we seldom miss a weekend purchasing all the fresh produce we can at the nearest farmers' market. However, as it is a ten-mile round trip in a three-thousand-pound car to collect perhaps thirty pounds of fruit and vegetables, I appreciate this also is a lifestyle indulgence with a high carbon footprint. The local supermarkets, where I also shop frequently, have most of their produce delivered from afar, but it is only a half mile from home, and I must visit them in any event for supplies not sold at the farmers' market. Quite possibly this option offers a better carbon footprint for my summer fruit and vegetable purchases than our shopping at the farmers' market.

Other than personal enjoyment of such food shopping and getting very fresh produce in season, I question if the claims made by the dedicated adherents to the hundred-mile diet are realistic. For starters, the food mile concept is a bit of a misdirected concern, because transport only accounts for about 10 percent of the carbon footprint of food on the fork.[95] The basic production of the food, processing, storage, and cooking are much more significant as far as energy use is concerned. Cooking in the kitchen as the last energy-intensive step for food before the fork accounts for about two and a half times the typical carbon footprint typically allotted to transport.

Besides transport, a perceived advantage of the hundred-mile diet, even if relatively insignificant, is that it is better to support local farmers than more distant ones. I agree for those farmers who make the time and effort to sell directly to the public, whether directly off their farm, in a farmers' market, or both. For most farmers, direct sales of their produce is not viable because this is a very labor-intensive activity and works only if there is a family member with the time and interest in such an activity to make this a practical undertaking. I can recall in my youth when urban folks we hardly knew occasionally arrived at our farm expecting to buy chickens, eggs, or perhaps a lamb. This was disruptive to our work schedule, and my father would politely explain that he was a farmer and not a grocer, usually sending them on their way empty-handed.

In the case of Ottawa where I live, there are close to a million people residing within the metropolitan area. Although agriculture within city limits is the third largest economic activity after government and the high-tech sector, there are only sixty farmers listed in the "Buy Local Food Guide" who will sell directly from their property. It should also be mentioned that many of these are located outside the city limits and require driving some distance to their properties.

Except for these exceptional folks, who mostly have product to sell only in the summer months, the overwhelming majority of farmers prefer to sell to food brokers, with their product going to the anonymous public. In such instances the proximity of these consumers is of little concern to the typical farmer. Conversely, why should a consumer prefer for their food to come from a farmer they will never know eighty miles away compared to another grower who is equally anonymous a thousand miles distant?

Even the freshness issue does not always hold true with local. Take a southern hemisphere apple that, after delivery by ship, with an exceptionally low carbon footprint per mile traveled, arrives at the market in North America in May within a month after it was picked. At this time local apples will have been stored for up to nine months as the entire North American harvest takes place early in the autumn. It is possible that the carbon footprint resulting from the climate-controlled storage over winter corresponds to the amount of carbon expended to deliver the more distant product. Precise comparisons of the two carbon footprints are difficult to find, but logic dictates that local apples may not always come out ahead in this regard. However, freshness of the southern hemisphere apples would be difficult to dispute.

Another disruptive element regarding eating local is the indisputable fact that North America is heavily urbanized. There is little opportunity for the population of larger cities particularly, on the coast with only one side to draw food from, to even come close to self-sufficiency within a hundred miles.

As stated previously it is sensible for consumers to purchase local food products wherever they can, and efforts to expand urban agriculture should be encouraged near major cities, but to consider such endeavors as a solution for community food security is not realistic. The large family farms and owner-operated ranches in the Great Plains and elsewhere, where there are expanses of fertile land and modest populations, are absolutely essential for massive quantities of sustainable domestic and global food supplies.

The hundred-mile diet adherents, of which I am a casual player, are serving themselves well in their quest to become better connected with their food supply. They also serve to provide smaller local farmers with the opportunity to sell their produce, dairy, eggs, honey, or whatever is in season at retail prices, vitalizing this segment of the farming industry. However, as most food production takes place in remote parts of the continent, in many instances hundreds of miles from a population center, this movement can at best provide only a fraction of the food needs of any large population center, particularly where winter is a factor.

There is also the belief by some that local foods are somehow better than their counterparts from afar. In reality every pound of food that is produced on this planet starts out as local to those who live nearby. Take Manhattan, Kansas, and their local wheat for example. There seems little reason for those in the other Manhattan to concern themselves about the quality of this wheat just because it is no longer local. Freshness will not be an issue as the wheat harvest in both locations will be approximately at the same time and grains store almost indefinitely without noticeable deterioration.

In spite of the best intentions of those who are committed to the hundred-mile diet, the mismatch of the geography of food production and centers of consumption do not lend themselves to making this more than an interesting and highly visible niche in the overall food supply dynamic.

Take Canada, which has a population of approximately thirty-three million which represents about one half of one percent of all humanity but produces 1.5 percent of the global food supply.[96] In other words the country produces roughly three times the amount of food it consumes, sufficient to feed another sixty-million-plus people, or approximately the population of France. Most of the agricultural output that is surplus to the needs of Canadians is produced in the three Prairie Provinces of Alberta, Saskatchewan, and Manitoba, a relatively remote place as far as world populations are concerned. To the north there is tundra and the Arctic Ocean, to the west there is British

Columbia with a modest population and then the Pacific Ocean, and to the east there is the rest of Canada (which is food self-sufficient) and the Atlantic, and finally to the south there is the United States that is an abundant producer of the same food products. In this situation it is difficult to think local when so much food is efficiently produced thousands of miles from the nearest overseas market. This same principle of major production regions far from populations requiring the food also applies to Australia, New Zealand, Argentina, and the United States.

To summarize, the 100 mile movement is a good one wherever practical and can work reasonably well at the peak of the growing season. However, to avoid food items simply because they come from afar achieves little regarding the overall efficient movement of food and only seems to preclude the enjoyment of the bounty and variety of the food that is available thanks to advanced storage and transportation technology and infrastructure.

12.

Pros and Cons of Knowing Your Farmer

During the summer of 2008, my wife, three of our adult children, and I rented a stately old farm house in Poland near the Czech border. During our one-week stay, four local farmers arrived randomly at the door selling wild berries, forest mushrooms, home-made cheese, free range eggs, and freshly baked bread. I speak German, and as this was in the area of Poland that was formerly under German control until after the Second World War, these folks still had enough of the language to carry on a basic conversation.

This clearly was a good situation for both the farmers and for my family. We received food that was exceptionally fresh and delicious, that was local and essentially organic because that was just the way they were grown but without special effort. The prices compared favorably to the shops but were much higher for the farmer than they could have received if selling wholesale at the farm gate. Furthermore, the farmer seemed to enjoy the banter and hearing about farming in North America.

Surprisingly there is a downside to this seemingly idyllic situation. For starters, for those who care about the carbon footprint associated with food production, this is a very inefficient method of distributing groceries. From what I could gather, each spent a half day driving his small truck over twenty miles to sell a few dozen pounds of food. The gasoline consumed for this probably more than offsets the carbon footprint disadvantage of eating food that was not grown nearby. Furthermore, a farmer is a highly skilled individual with land, machinery, and other assets to produce food. On the other hand, the skill set required to sell food is much easier to attain. Thus in purely economic terms of the best use of resources to produce food, local or otherwise, farmers are best left on their land and could leave selling their food to others. To draw an analogy, this is somewhat like asking the family doctor to run the checkout at the drug store. Yes, it would be comforting to have this expertise at the point of purchase of one's medication, to provide the final assurance that everything was in order. However, this would clearly represent a squandering of talent that is badly needed elsewhere.

In spite of the above reasoning, I indulge in purchasing directly from farmers whenever I have the chance, as I did with my wife a year later when we rented a two-hundred-year-old house in a quaint village in Crete. This time it was only yogurt plus cheese from one farmer, with little chance to communicate, but I still enjoyed being close to the source of my food. However, does the farmer really want to know the consumer? I would venture that the answer is most often no. As I stated in the previous chapter, my father did not appreciate it when either strangers or acquaintances came out to the farm to buy whatever we might have available.

Thus in early 2010, when the US secretary of agriculture announced the program "Know Your Farmer, Know Your Food," commercial farmers probably rolled their eyes. A friend of mine at the National Association of Wheat Growers in Washington DC was quite amused by this idea and the disconnect with most of their members, many with thousands of acres and farming a few hours' drive from any major population center.

If you listen carefully to the secretary's message, it is somewhat more relevant and broader in scope than the program's sound bite might suggest. After decades of relative indifference by the food-consuming public to where their food comes from, the USDA was reaching out to this newfound and increasingly savvy movement to facilitate a meaningful flow of information about food and food production to the public. Also, the intent is to provide possible new opportunities to the million or so smaller farmers who may have the time with their limited food-producing asset base to effectively market the fruits of their labors directly and at retail prices to a willing consuming public. However, it is very doubtful that many commercial scale farmers will be active participants in this program because they are not set up to accommodate strangers while they are busy producing food.

There are some who sincerely think farmers' markets and active farmer participation in the direct marketing of food is the agriculture of the future. A few months ago I attended a local meeting arranged by those who had a strong interest in current food trends. There were three speakers on a panel with a Q&A session following their presentations. One presenter was head of the local farmers' market association, another was a local organic farmer, and the third owned and operated a restaurant that featured local organic foods. All were very pleasant people who strongly believed in a close relationship between farmer and consumer.

From their individual points of view, they all had a good business model that worked well for them. The presentations were interesting, but their credibility took a hit for me when they went from describing their business to professing they were of the opinion that they were on the vanguard of a movement that would eventually displace commercial agriculture and large-scale grocery chains. Yes, they served their niche well, catering to prosperous individuals with a focused interest in food, but to consider this a model for feeding a public much beyond who they were serving now was at best naïve or possibly outright disruptive to our overall food supply. What they were proposing was a labor-intensive approach that has the potential to supply less food for a hungry world.

In summary, knowing your farmer is perhaps shorthand for understanding the whole dynamic of the food industry, of which farmer's play a vital role. It is my intent that this book will help readers in this regard.

13.

Legacy, Heritage, and Heirloom Varieties

There is a fairly widespread lament that thousands of heritage fruits and vegetable varieties have disappeared because of modern agriculture. Reportedly there were an estimated four thousand varieties of heritage apples and another two thousand of heirloom tomatoes, with similarly large numbers of legacy potato varieties and other fruit and vegetable choices available to small-scale farmers and gardeners earlier in the last century. Is this reality or partially an urban legend, and is today's consumer of fresh fruits and vegetables really less well served regarding choice than were their ancestors?

While the nation-wide variety of fruits, vegetables, livestock, and poultry that existed decades ago was undoubtedly higher than today, how great is our loss in reality? Let us look at the very high numbers of all the varieties that supposedly existed in these happier times. Where there really this many options, or perhaps a multitude of local names for similar or identical fruit and vegetable types? Starting with my great grandparents and their four European languages, each had a different translation for apple: apfel (German), jabiko (Czech), aeble (Danish), and eple (Norwegian). And then most probably each have had, for example, different names for earlier versions of the variety "delicious," which, if translated literally, comes out in each language respectively as koestlich, chutne, laekker, and deilig. Perhaps names have disappeared but probably not the actual heritage fruit or vegetable. Also, do very slight differences in one heritage tomato or apple, when compared with another, really count as a distinct variety that matters to the typical consumer? In other words, is the loss as staggering as the very large number of reportedly lost varieties implies?

A quick check of heritage tomato availability on the Internet locates one nursery in California offering over 160 varieties of heirloom tomato seeds alone. Another outlet has seven hundred heritage varieties of about a half dozen common vegetables. At the farm supply center a short drive from my home, I counted nineteen varieties of seed potatoes available to both the home gardener and the commercial potato farmer. These numbers are far from thousands of varieties, but they do represent a reasonable choice of probably all the main tomato and potato types that existed three or so generations ago. As few, if any, individual gardeners have the time or patience to try everything that is out there, we are not being deprived of adequate choice.

There is nothing like this number of varieties available even in the best stocked farmers' market of today, and even more so at the supermarket. However, anyone who has the slightest inkling of the concept of inventory control will be well aware that such choice of hundreds of apples or whatever at any one outlet is as impossible now and as it was in the past. Today the choice is actually quite impressive and probably rivals or betters that which my ancestors enjoyed. At the supermarket closest to my home one winter day, I counted ten tomato types with colors, including bright red, yellow, orange, and brown. The sizes varied from that of a small grape to monster field tomatoes of a pound each. This was a greater opportunity for choice than I needed. In the best of times for heirloom tomatoes in northern climes, they were only available from the garden or in the local market from midsummer until the first frosts in autumn, perhaps three months at best. My winter choices would have been unthinkable to my ancestors who would not have seen a tomato for many months of the year.

The point here is the real choice of fresh fruit and vegetables in the marketplace that we have today for the entire twelve months of the year is far superior to the assumed "golden era" of the heirloom fruits and vegetables, which were both seasonal and regional when it came to availability to consumers.

No one would deny that a freshly picked heirloom tomato is a tastier product than one finds in the supermarket. I know this from practical experience from the unique varieties that I grow in my own garden. However, their downside is that most such varieties do not yield or store particularly well, and this means that the modern food industry must give these a pass. I maintain that they were never intended for commerce but were great for the home gardener where they are still to be found today.

The commercial tomato producer, organic or otherwise, is asked to provide a product during the winter that looks reasonably freshly picked and is not too expensive for consumers living perhaps a couple thousand of miles from the field in which it was grown. The obvious loss is a ripe and ready-to-eat tomato available for consumption at the time of picking. Instead harvest must take place before fully ripe to ensure the product has both the shelf life and is firm enough for transport. Also, varieties are selected that mature on the vine at the same time to facilitate an efficient once-over for the field harvest but are naturally firm and also yield well. Unfortunately flavor, while still a quality that growers attempt to provide, may suffer in the process.

The point here is that those who want tomatoes all year long must rely on distant fields and a product that will withstand the stress of transport and storage. I maintain that for most consumers such a tomato is better than the dried or canned version, which was the only offseason choice of our ancestors for much of the year, unless they lived in a very benign climate.

14.

So You Think You Can Garden!

As readers will have gathered, I am an avid gardener and share a wonderful hobby with the dividends enjoyed by countless millions of others. There are few other activities that have so many benefits, including exercise, fresh air, closeness to nature, and the bounty of fresh fruits and vegetables that bring nutrition and culinary delights to the business end of the fork. In essence, a garden is a small farm.

Like farming, gardening is often a challenge, and as all gardeners know there are failures and partial success along with great yields. Through experience the results improve, but less than stellar harvests are to be expected with some regularity. Thus when writers or commentators encourage urban dwellers to rip out their lawns and replace them with a vegetable garden and enjoy the fruits of their labors with little mention of the effort, there is the possibility of disappointing results. And because there are only relatively short periods when harvest is possible which means reliance on the supermarket most of the year, I sometimes wonder if the people making these recommendations are actual gardeners themselves.

I am a gardener who was raised on a farm where we had a large but very seasonal garden, which launched me on my lifetime passion for producing my own fruit and vegetables wherever practical. Friends and neighbors often ask for my advice, so I seem to be reasonably informed and experienced compared to others. I am also blessed with three-quarters of an acre of urban property. There are old growth cedar and maple surrounding my gardening area, with a few maturing maples throughout the property. Here lies my first challenge: trees and their shade.

When I purchased my home over twenty years ago, I acquired the utensils required to tap fifteen of the maples for syrup, and I replaced my lawns with hostas or perennials where shade was an issue. I plant two apple trees, two pear trees, a cherry tree, male and female winter hardy kiwis, four hazelnut bushes, and a raspberry patch, and I also prepared two reasonably sunny plots for vegetable gardens.

The maple syrup harvest in March the first year was a resounding success, with over a two-gallon yield from eighty gallons, or nearly two barrels of sap. With snow still on the ground, what a thrill to be taking in the first harvest of the year! Over ten or so days, during my hours between sleeping and work, two pots constantly boiled on our kitchen stove. By the end of the season my wife noticed that sugar crystals were on many of the walls and ceilings throughout the house, which required substantial washing to make right again. Worse, our middle-aged kitchen stove developed burner fatigue to the point that it had to be replaced. Factoring in the electricity costs, a new stove, and an unhappy wife, my sugar bush activities have been scaled back to a token harvest of perhaps one-tenth of the potential now, where boiling is carried on outside in the early spring cold . No crop failure here, but technical challenges preclude a full harvest. Incidentally my maple syrup is the ultimate organic product, as the soil under the trees has never been disturbed by humans.

As for my other gardening activities, my first challenge was the same maples, which are the top of the flora food chain in our area and dominate all other plants on three sides of our property. These plus the ones in the yard also grew robustly and jointly reduced the available sun to the point that the cherry tree, the two pear trees, and one apple tree died. One apple tree remained, but without the potential to cross pollinate from another, it bares no fruit. While the hazelnut bushes survived, they also do not produce, presumably because of the lack of sufficient sunlight. Also, one of my two vegetable garden patches faced the same sunshine-deprived fate and fell out of production.

As for the kiwis, the garden center where I purchased them advised me to acquire both male and female plants so they could cross pollinate and produce fruit. I ended up with a gender mix-up, and although I have a monster pair of vines and plenty of tiny blossoms in the spring but there is no fruit as I have either two female or two male plants, I have since purchased a self pollinating kiwi variety and have a modest harvest of grape sized fruit.

My next challenge is that I share my garden with rabbits, groundhogs, gray, black and red squirrels, chipmunks, and at least twenty five species of birds. The raspberries were the first harvest that is surrendered to the birds every year, with only the occasional handful left over for us. Most vegetables are similarly impacted by my furry friends, except tomatoes. The rodents may be leery about this plant because its cousin, the deadly nightshade, is endemic and can be found on the margins of the property.

Because of the success with tomatoes and my ever declining sun-blessed space, that is now my only real crop, but I am blessed with a bountiful harvest from mid-August and well into the fall. I share the fruits of my labors with neighbors, plus I freeze a fair quantity as tomato sauce. Some years I am also able to extend the season a couple weeks for vine-ripened fruit with my late-producing varieties by pulling the unharvested vines before the first frost and hanging them upside down in my basement.

In spite of all this effort and early promise of bountiful harvests from spring to winter, basically all I have to show for my effort is a token bottle of maple syrup and a few weeks' ample supply of tomatoes. Though my challenges may be extreme, many gardeners will share similar experiences of disappointing results for a host of reasons. Relatively short growing seasons in many parts of the United States and all of Canada where winter is the greatest impediment to feeding the family from the garden except for a few months.

Readers might be wondering what my gardening experience has to do with demystifying the passage of food from farm to fork. I am attempting to provide a reality check on the potential impact that home gardens might have regarding our overall supply of food. Some well-meaning folks actually think that home gardens are another answer in food production that will replace a significant proportion of commercial agriculture.

For starters, the amount of land available for growing food in urban areas is actually quite insignificant when compared to the areas already dedicated to farming. As readers will recall from the chapter on farming and biodiversity, the US landmass is 2.1 percent urban while 19.5 percent is classified as farmland (this does not include pastures/grasslands for grazing cattle and sheep). Thus the urban area total is one tenth that of land dedicated only to food production. Furthermore most larger urban areas were originally established because they were a natural transportation hub either on water or a railway junction. Thus it is only by chance that most of the available land in many cities is ideal for farming on any scale. Then, of course, there are all the roads, homes, apartment buildings, offices, public buildings, factories, airports, railway tracks, and much more which preclude any type of food production in an urban setting. While perhaps not a major factor, some available land in cities may also be contaminated and thus rule out this activity.

While roof top gardens are touted as available space and already are occasionally in place for cosmetic purposes, they require immense amounts of water (probably drawn from the regular city supply) because of high evaporation rates as there is no subsoil that is essential to retain moisture. Also the soil on roof tops must be "borrowed" from farms somewhere so this is not a particularly sustainable practice. This basically leaves lawns as the prime space available for urban food production. As a guess, perhaps ten percent of an urban setting is covered in grass or is otherwise suitable. Given that farmland used only for food production is ten times that of the total urban area, this means that the available space for food production within cities might be one percent of that already dedicated to the cause in rural settings. Thus the best case scenario if all lawn owners turned these into mini farms, and also had the inclination along with skills to seriously produce food, the total US agricultural output might increase by one percent. Thus I maintain that while gardening is an outstanding hobby it is hardly a viable activity to make the impact on food security that some who promote the concept and indeed some politicians make it out to be.

For example, a candidate for mayor of Calgary, Alberta, a city with a population of a million, ran partially on a platform with a focus on producing 50 percent of the food requirements of the city through urban gardening and backyard chickens. He seems to have overlooked the fact that his city is several hundred miles north of Fargo, North Dakota; is of a much higher elevation, and has a frost-free period of only 120 days. This climate precludes any significant fruit production, and vegetables will be in season for about two months. Furthermore, summers tend to be somewhat lacking in precipitation, as is evident by the fact that trees do not grow there naturally except in river valleys, so gardens will need to be watered from the urban water supply if they are expected to thrive. This perhaps is a somewhat costly proposition to the homeowner, but more importantly, it will tax a system which is not designed to support widespread irrigation. In other words there is no chance that this effort will provide more than a very small portion of the vegetable requirements beyond the late summer and early autumn period. He also failed to consider that beyond the existing gardeners, most of the unsuspecting population, probably with little interest or inkling on the finer points of gardening, will have to be mobilized to garden in places where no food is grown today.

This was a lofty goal that after an election campaign was probably quickly forgotten, and it is quite harmless as a passing issue. However, if there are others such as this candidate who are of the opinion that the farm-to-fork system that is in existence today is broken and that urban gardens are a viable option capable of replacing much of the farming as we know it today, this could become problematic. Growing food within a city is a great initiative and should be encouraged, but at the same time conventional agriculture will continue to provide most of our food. Though I doubt if it will ever happen, any initiatives to develop urban gardening at the expense of the existing food production is not a realistic option.

15.

Weather, Climate, and Food

What happens regarding food production on the farm, well before the journey to the fork, is very much influenced by a combination of climate and weather.

Climate determines what crops can be produced in a particular geographic region; weather, the annual variable, establishes how much will be produced in any given year.

In this context, climate is defined by the combined long-term averages of precipitation, the hours of sunshine/overcast, temperature, and, as one moves further from the equator, the frost-free period. Weather is the variation in all of these in any given growing season, which might be unusually wet or dry, cold or hot, unusually sunny or overcast. A season can have an unexpected frost at the beginning or end of the growing season, plus storms with hail or high winds. Any of these on their own, when extreme, can devastate a crop. However, all these factors must align themselves favorably each growing season to result in an abundant harvest.

Given that there are so many variables in weather, it is approaching a miracle that for enough of the food-producing world each year the rains come when they should, severe storms are few and far between, northern and southern climes distant from the equator are spared killing frosts when the crop is either just getting established or in the later stages of maturity, the heat units are sufficient for crops such as corn but not too hot for more temperate crops and grasses. Although farmers cannot do a lot about it, they are acutely aware of all these variables and what their impact is on their fields and pastures on any given day of the growing season.

Farmers have little control of the weather, but they generally manage nicely around climate, in particular through the judicious selection of which crops to grow. To use an obvious example, conditions are impossible for oranges to grow in Iowa. There are also much more refined crop selection options, such as the varieties of wheat that thrive in Texas but will not do well in the Canadian Prairies, and vice versa. The different traits in each wheat variety mitigate climate diversity, thus enabling regular bountiful harvests in quite dissimilar regions. Precipitation challenges of either too much or too little are alleviated or even eliminated through irrigation or tile drainage systems. When climate is not favorable for grain or other crops, then the grazing cattle, sheep, and goats remain an option where sparse grass still yields some food on every acre.

Thanks to centuries of plant breeding, including recent genetic modification, there is an appropriate selection of crops to match what nature has to offer in a very wide range of climates. For example, the frost-free period is crucial for the selection of the crops that will mature within a very specific period of time. Although farmers and gardeners are very familiar with the concept, the frost-free period in any location is the average last day in the spring when the temperature does not go below 29 degrees Fahrenheit (three degrees of frost, which will not kill most plants) and the corresponding average date in the fall. "Frost free" is a bit of a misnomer because there is a fifty-fifty chance that there will be a frost either after or before the average spring and fall dates. This is a twice yearly unknown of critical importance to farmers and gardeners, who must select the seeding dates and the varieties that they grow to match the realistic dates when frosts are unlikely.

One of the most northerly farming communities of consequence in North America is Peace River, Alberta, with a frost-free period of ninety-nine days (May 26 to September 3).[97] Farmers here operate in a very narrow window of opportunity to plant, grow, and harvest their crops, and they manage by planting fast-maturing barley that can mature in a little over eighty days from the date of seeding. Early varieties of canola, peas, lentils, and mustard seed also fit this growing season. However, wheat, with a growing season of over a hundred days, may be a risky venture. Nevertheless farmers in this northern region still have an impressive selection of valuable foods crops that they can grow successfully.

Moving south, the frost-free period gradually increases with significant variations for elevation, with plateaus and foothills obviously experiencing shorter growing seasons. For example, Aspen, Colorado, with an elevation approaching eight thousand feet, has a frost-free period of only eighty-eight days.[98] At a comparable latitude, for the Midwest Great Plains the frost free period is typically from mid-April to late October, or approaching two hundred days, which is more than ample for most annual grain crops. Here temperatures, or heat units, are particularly critical to corn and soybean production. This is based on the concept that these crops need a certain aggregate number of reasonably high-temperature days to facilitate growth and maturity, only thriving when the accumulated number of hours over the growing season is at or somewhat above this level.

This concept is particularly obvious as one drives north on the I-29 from Sioux City, Iowa, through South and North Dakota along the Red River and across the Canadian border into Manitoba. In northern Iowa and much of South Dakota, corn and, to a slightly lesser extent, soybeans dominate. Then north of the border between the two Dakotas, fields of wheat and sugar beets become more and more frequent until one arrives in Canada, where large-scale commercial corn production is still relatively uncommon. And yet thanks to aggressive corn-breeding programs, the Corn Belt is steadily moving northward, primarily at the expense of wheat production. New varieties require less heat units and perhaps less frost-free days and thus can now thrive in what once was an inhospitable climate for this crop. This is a perfect example of farmers, along with the essential help of plant breeders, managing climate.

On successfully meeting precipitation challenges, the Central Valley of California, as well as southern Idaho and many other places in the world, would essentially be deserts if it not for the fact that the melting snow pack that accumulates in the mountains over the winter is captured each spring. Not only do farmers use the water to make the desert bloom, but particularly in Idaho, the system of dams to store the water also provides essential downriver flood control each spring.

Alternatively, when excessive rains are the climatic norm, and the topography is reasonably flat (as is the case in much of the Midwest), subsurface drainage effectively removes surplus water through a system of tiles buried beneath the surface at regularly spaced intervals. This system lets gravity do the work by taking advantage of the natural slope of the land. In drier regions moisture is conserved through advanced tillage practices that leave the soil virtually undisturbed when planting, and it also provides a residue from the previous crop on the surface. Both outcomes of this very limited tillage substantially reduces evaporation and leaves more badly needed moisture for whatever crop the farmer has decided to plant. It is through all these water-management practices that otherwise fertile land would not yield the vast quantities of food that are regularly harvested not only in North America but throughout the world.

Perhaps the most uniquely favored country in the world, with an ideal combination of climate and weather suited for the production of food, is New Zealand. I lived four years in Wellington, where it never froze once at street level and seldom got over 80 degrees in the summer. Regarding climate, the rainfall averages about sixty inches, which falls evenly all year over much of the country, so there is no excessively wet and or dry season. With constantly mild temperatures and regular rainfall twelve months of the year, the grass for grazing sheep and cattle is perpetually lush in at least most of the North Island. Even in the South Island, except in the extreme south where it does snow, grazing is a year-round activity.

As an aside, it may appear that farmers in New Zealand require little effort compared to elsewhere to meet climatic and weather challenges, but their presence in the somewhat remote South Pacific has an interesting history. For both the British home front and to meet the needs of an extensive empire, the food requirements were beyond what the mother country could provide. Given the superb weather and climatic conditions, New Zealand was established effectively as an offshore food producing province of Mother England, which dispatched some of their brightest and best farmers to establish extensive grazing and dairying operations. To this day New Zealanders will remind their Australian neighbors of the fact that their ancestors were the landed gentry while across the Tasman Sea; Australia was the dumping ground for prisoners. (I hasten to add that somehow the convicts got their act together and have long since become a nation of outstanding farmers.)

I often marvel that globally, farmers of all types somehow produce enough food to feed the world year after year. Thought crop failures occur either locally of even at the regional level every year, the entire food system has sufficient diversity to consistently provide enough to feed nearly seven billion people year after year. It is often overlooked that this virtually perpetual horn of plenty functions in an extremely narrow range of climate and weather conditions.

Take this Goldilocks, "not too hot but not too cold" range of temperatures that plants need not only to survive but also thrive if ample food is to be the outcome. The freezing point of 32 degrees is generally considered the point at which plants can no longer flourish, and many will perish. At the other end of the temperature spectrum, 110 degrees for periods beyond a few hours is another threshold where many plants succumb and few thrive. In reality, most plants –and indeed animals, birds, and insects –do not function well outside the temperature range of 50–90 Fahrenheit, a mere forty degrees. When one considers that the temperature in the solar system varies, from the surface of the sun to the nighttime surface of the moon, it is incredible that this 40 degree temperature range prevails over much of the earth for many days of the year and has done so for millennia. I find it amazing that we live on a planet where living creatures and plants only do well in a very constant temperature range that is less than 0.5 percent of the extremes of our solar system.

I appreciate that for evolutionists, it would be expected that both flora and fauna would develop over time in a way to best take advantage of existing climatic conditions, and this is what happened on our planet. However, it is difficult to imagine any sophisticated life-forms developing outside of the 180degree range between the freezing and boiling points for water, which still represents a miniscule 2 percent of the total temperature variation found on our nearby solar neighbors. My point here is that the most abundant substance on the Earth, water, can only support life when it is in liquid form and not as a vapor or a solid.

This constant, narrow temperature band is the result of a very complex dynamic. For starters, the Earth's orbit around the sun is nearly a perfect circle. If overly elliptical, there would be significant variations in the distance from the sun with much more pronounced hot and cold extremes, temperatures well outside this narrow comfort zone. Secondly, the distance of the sun from the earth is consistently very close to ninety-three million miles and is precisely where our heat source should be to achieve this perfect temperature scenario Furthermore, the Earth's atmosphere forms a "blanket" which traps the heat and keeps temperatures from freezing at night, particularly in temperate regions from spring to fall.

A fourth factor adding to this perfect dynamic is the twenty-four-hour rotation of the earth on its axis. If much slower (meaning a longer day), the nights would be colder, the days would be hotter, and the precious temperature range enjoyed by the Earth's inhabitants would be severely compromised, thus precluding the life we know on this planet.

Another essential element is the temperature-moderating effects of the oceans that cover much of the earth. The oceans, along with the dynamic of evaporation and then condensation at higher and colder atmospheric levels, provide the precious rain and snow that falls on the land.

Thus to have this very narrow range of temperatures, we enjoy a perfect distance from the sun, an ideal twenty-four-hour day, a circular orbit around the sun, an atmosphere that filters the sun's ultraviolet rays and impacts on the heat that reaches the surface and remains over night, and vast areas of open water to moderate temperature swings plus provide the moisture for precipitation. Added to this complex set of essential conditions, the axis of the earth is tilted somewhat so that as it rotates around the sun, we have the four seasons in northern and southern regions but near constant tropical temperature conditions at the equator. With variable climatic conditions, the opportunity exists to have a much greater diversity of plants and animals suitable for food and other resources.

Then there is the amazing process of photosynthesis, whereby plants use their leaves to convert the nutrients, primarily from the soil, gases from the atmosphere, and the energy from the sun into a host of edible and useful materials for not only mankind but all creatures. Interestingly, thanks to photosynthesis we have not only massive reserves of coal from the accumulation of plant material millions of years ago, but also crude oil and natural gas from other forms of life that thrived on the abundant array of vegetation. These ancient deposits of energy are a huge inheritance for humans, though unfortunately we may be squandering it to some extent today.

If any one of these many factors was not in place in a very consistent and precise manner, the bounty and variety of food and living conditions for all creatures would not exist. I sometimes wonder why priests, pastors, rabbis, imams, and other spiritual leaders do not expound more on this phenomenal combination of life-giving conditions to convince skeptical members of their flocks that there truly is a creator. For our daily bread and much more, humans owe so much to these unique fundamentals of our planet that provide the climate and weather that make a quality life for so many possible.

16.

Water and Food Production

Most of the water used for food production falls directly on the land where it is needed in the form of rain or snow, and its use for this purpose is generally considered to be sustainable and environmentally benign. Where the amount of precipitation does not match the potential food output of the soil, the possibility for irrigation exists, which usually comes from groundwater (wells) or a dam on a natural waterway. Within his rather well-known water-food dynamic there is substantial variability on the assorted water application practices.

According to the US Geological Service,[99] irrigation for the production of food accounts for 65 percent of all the fresh water (other than that which falls as precipitation) used in the nation, or 137 billion gallons per day. Clearly farming is the water guzzler, but before this is painted as a negative, it is useful to realize that from the sustainability perspective, much is available for the taking without competition from other users, and quite possibly with other important benefits beyond food production.

As a positive example of water usage for food production in the United States, take the Snake River in Idaho, a major tributary of the Columbia River that is the fourth largest river system in the country, draining an area roughly the size of France. [100] On the upper reaches of the Snake, in the mountains of Idaho, there are six major dams that capture water, primarily from the spring snow pack melt, and together produce nearly four thousand megawatts of carbon-free electricity, as well as providing badly needed flood control, particularly in the lower Columbia as it nears the Pacific. In addition, these reservoirs also offer great recreational opportunities. Then there is the added food dividend through the judicious release of water over the growing season to irrigate 1.1 million acres of rich but formerly unproductive volcanic soil that was close to being a desert before the water arrived.

The amounts of water applied for irrigation are carefully controlled, with fish and the healthy environment of the two river systems sharing a priority with agriculture. In this application the water supply is sustainable, and agriculture is but one of four beneficiaries of the system of dams.

As for the farmers, they benefit from abundant fields of alfalfa, potatoes, wheat, malting barley, and other crops. It should also be mentioned that farmers only receive water that is surplus, above what is required for the environmental well-being of the river system. In this regard they have a basic coping strategy when the available water is challenged: they utilize the rationed supply on the higher-value annual crops and leave their perennial alfalfa fields without irrigation, with the knowledge that the deep roots of this perennial crop will respond immediately when water is again available.

Irrigation of a Grain Field

Environmentalists and others who are rightfully concerned about the immense amounts of water used to produce food tend to treat all such usage as more or less equally environmentally challenged. I suggest that they should take a careful look at the entire dynamic of the application and then focus on where the water draw for food production is not sustainable or is not the best use of a precious resource.

The entire Columbia River Basin, which originates far into British

Columbia, is managed in a coordinated manner between the two countries through the 1964 Columbia River Treaty.[101] and the treaty is a good example of the sensible utilization of a water resource that straddles an international border. Overall, the multiple benefits that this highly developed water system provides are generally considered to far outweigh any negative consequences.

An entirely different situation of an unsustainable misuse of fresh water is exemplified in the recent history of the Aral Sea. In the 1960s, in an ill advised effort to boost cotton production in the region, the former Soviet Union diverted much of the flow of the two largest rivers that emptied into what once was the world's fourth largest saline sea. The results for the Aral were disastrous, with the water level dropping by over sixty feet between 1960 and 2000 and the volume falling to 20 percent of the pre-irrigation level. This is arguably the most extreme incidence of environmentally unsustainable water usage for agricultural purposes.

Back in the United States, a well-known example of an unsustainable use of water for food production is the Ogallala, or High Plains Aquifer, which according to the US Geological Survey[102] (USGS) provides about 30 percent of all groundwater used in the country for food production. It is applied to 27 percent of the irrigated cropland and supports nearly one-fifth of the national production of corn, wheat, cotton, and cattle This massive aquifer lies under parts of eight states, from Texas to South Dakota, and since significant pumping began in the 1940s, depletion is increasingly a problem in some areas, with water table reductions of up to a hundred feet. This unsustainable practice of mining this resource is analogous to drawing funds from a bank account without any means or prospect of replenishing them.

Not only is a resource being depleted, but the costs of incremental energy required and more expensive equipment to pump the water from ever greater depths is another issue. Thus as the water table continues to recede, there is the possibility that wells in some locations will cease to be economically productive and will be abandoned.

Although land subsidence or sinking does not seem to be as serious a problem over the Great Plains Aquifer, it is an issue with lesser but still significant aquifers elsewhere in the United States where the surface has subsided by ten feet or more. Also, as groundwater below sea level is frequently saline, even a considerable distance inland, the depletion of the freshwater may cause salty water to enter the irrigation wells with severe deleterious effects on the crops and soil. Once contaminated, such wells often must be abandoned, and for the farmland above, irrigation may no longer be an option, with a significant drop in yields and property values.

Of the sustainability and environmental issues facing US agriculture, groundwater depletion in several regions is probably the most significant challenge.[103] In a discussion with aquifer-dependent farmers in Nebraska, there is no doubt that the issues are well understood and appreciated. Considering natural recharging occurs at only about one inch per year, the resource will eventually run out in many farming areas. Outright banning is not considered an option because leaving the water underground does little good, as there is no recreational or other benefit for this resource. The possible solution seems to be establishing a rationing regime that will encourage farmers to adopt more efficient irrigation technologies and produce less thirsty crops. Though depletion will still be a factor, it will occur at a substantially reduced rate.

On a different topic related to water use in agriculture, there is substantial attention paid to how much is required to produce a pound of a variety of foods, with beef often portrayed as the most wasteful because it requires up to 2,500 gallons per pound, according to some observers. The beef industry estimates that the amount is less than a fifth, but even so, this is approaching 500 gallons, which is more than a person drinks in a year[104].

This seems to be a lot of water no matter how one does the calculations, but what seems to be overlooked in the debate is that water used in agriculture comes at varying degrees of costs to the environment. To illustrate the point, beef in New Zealand is grass fed on pastures that, as previously mentioned, receive adequate rainfall twelve months of the year. If one calculated all the water that fell on the land, the amount of water to produce each pound of edible beef would be quite astronomical. However, so what? If all those millions of acres were abandoned and there was no grazing, the rain would still fall, and vegetation would continue to grow, but what would be the benefits of avoiding the agricultural use of water when the downside would be less top-quality animal protein available to feed the world?

Similarly in Idaho, where irrigation water is essentially a byproduct of flood control, recreation, and carbon-free electrical generation, its use to produce food, including beef, involves few if any negative sustainability or environmental issues . Move the beef production where the animal is completely dependent on the water for all its food and well-being from a rapidly depleting aquifer, and the issue becomes real. Thus it is important to realize that water usage for food production is often of little concern, and to simply calculate the amount required for a pound of beef or whatever and then express alarm can be a meaningless exercise, unless a challenged source of the water enters the equation.

Another major issue relating to water and agriculture is the possibility of capturing it and then moving it long distances to where it is in short supply and needed. Two North American scenarios that lend themselves to this concept are to keep deserts blooming in southwestern America and also to give the Ogallala Aquifer a rest. Here is where Canadians become paranoid that our supposedly abundant water supplies are ready and waiting for our southern neighbors to take under some clause of the North American Free Trade Agreement, or other mechanism. My opinion is that even if Canada had a policy favoring bulk water exports, the economics of such a scheme would not work given that food production is expected to provide the bulk of the revenue to make such a development viable.

As stated at the beginning of this chapter, agriculture is by far the major water-consuming activity in the United States, and it only works because for this application it is virtually without cost to farmers, as it is either based on historic water rights agreements, or as in the case of Idaho, there are few if any other uses, and the water is largely paid for because of other important benefits. To illustrate the point, I refer to the 2011 published irrigation water rate to farmers for Merced, California,[105] which ranged from $18.25 per acre foot for nearby district water up to $42.50 for the same acre foot amount from further away. For reference, an acre foot is equivalent to flooding a football field 12 inches deep and amounts to about 325,000 gallons or what it would take to fill half an Olympic-sized swimming pool. With the $18.25 water rate, this equates to less than two cents per 325 gallons, an amount equivalent to filling about twenty gas tanks of a typical car.

2011 Water Pricing and Sales Information: Merced, CA	
WATER PRICE:	
In-District	$18.25 per acre foot
Sphere of Influence (SOI)	$42.50 per acre foot
Sphere of Influence (SOI Drought)	TBD by MID Board of Directors

The bulk transport of liquids over any distance can be by pipeline, where the main product is crude oil or more expensive petroleum products such as gasoline. Given that a barrel contains about 42 gallons, the 325 gallons represents close to eight barrels, meaning that the most expensive Merced water for irrigation has a delivered price of about a quarter penny a barrel. Assuming a value of $100 per barrel, crude oil, which presents a business case for a pipeline, is 40,000 times more valuable per gallon than irrigation water. The point here is that pipelines would not likely be an economically viable means of transport for water.

The other alternative is for river diversions and canals to move the quantities of water, which works extremely well in various western US irrigation districts. Here is where gravity and the natural topography of the land must be favorable and usually involves sources that are in mountainous regions with irrigation in the flat land below. A quick overview of the lay of the land in western North America leads one to realize that such a system does not work for Canadian exports of water to the United States. It may surprise some but Canada is generally downriver from the US.

In British Columbia there are two major south-flowing rivers, the Fraser and the Columbia. The Fraser, which is the more westerly, flows straight south from the interior of the province, and almost in sight of the forty-ninth parallel, it veers directly west and empties into the Pacific. Diversion of this river from close to sea level, through some rather challenged terrain, would be a major engineering feat and perhaps is not worthwhile, particularly as this corner of North America is already fairly well endowed with adequate rainfall and is also partially serviced by another river system, the Columbia.

The waters of the Columbia are already well managed under a bilateral treaty. Canada temporarily retains the spring runoff by a serious of dams producing hydroelectric power, and it uses some for irrigation, but thanks to gravity, it lets substantial quantities flow to the United States in a timely manner to meet water requirements, primarily for irrigation and other benefits. Things work well for both parties, but there is no additional water here for export.

From here it moves east into the relatively arid prairies, where to the west the water comes from the mountain snow pack and receding glaciers and flows north, either into either the Arctic Ocean or north east into the Hudson's Bay and on to the North Atlantic. Gravity works against water exports from this vast region, particularly as the rivers involved here are some distance from the US border. The only exception in the western prairies is the tiny Milk River in southern Alberta, which is the far northern headwaters of the Mississippi river system, with the water already flowing south and as the topography intended, thus offering no incremental opportunity for export.

Then for several hundred more miles east, there are no rivers of consequence near the border until the Red River, which flows directly north into Manitoba from North Dakota and causes serious flooding in Winnipeg and surrounding farms on a regular basis. If the Great Plains were not so flat and with such shallow river valleys, the reverse of the Columbia River Basin would be an ideal solution, with Canada encouraging the US to hold back spring runoff water for flood control, electrical generation, and diversion of a reasonable amount for irrigation. No potential for water exports here.

Further east a huge area of both countries drains into the Great Lakes and into the St. Laurence River, which flows northeast into Canada and on into the Atlantic. Like the Columbia River Basin, this is also a well-managed, bilaterally controlled water shed with both nations fully aware that this resource has little or no incremental water to offer to agriculture.

There is also another economic factor which further mitigates the chance that large-scale water exports from Canada to the United States are economically viable. Assuming that water exports were acceptable to Canadian voters, the likely scenario for such exports would be to somehow divert the rivers that tend to meander northward in the prairie region to the area of the United States to the Great Plains Aquifer and further to the southwest. However, while the growing season may be shorter in the Canadian prairie provinces, it does respond spectacularly well when irrigated for the production of a wide variety of grains, vegetables, potatoes, and other crops. As the economics to move the food to market always favors growing it near sources of available irrigation water, the same would apply in this situation.

Should it ever be feasible, instead of moving immense amounts of Canadian water into the United States for irrigation, the business case would be to use it as close as possible to its source. Given that it takes hundreds if not thousands of gallons of water to produce a pound of food, the transport of the food seems to be the more promising of the two options. However, with the receding glaciers in the Rockies, there is little likelihood that even the Canadian case to use these waters for irrigation has any future.

Also, the United States is actually water rich compared to most countries. Where the perceived shortage exists is the expectations of the never-ending and expanding flow of basically free water that so many agricultural communities depend upon for their bountiful crops. Conservation practices such as drip irrigation are being implemented to enable the same food output with substantially less water, but the fact that water is relatively abundant at close to zero cost compared to its true value inhibits progress regarding conservation of the resource. There is clearly a shortage of water at this very low cost, and this is where the perception arises of water scarcity, while counties such as those in the Middle East and China would view this as an abundant water supply in need of a bit of market economy dynamics.

Another issue that gives rise to what appears to be a shortage is any attempt to divert water that was traditionally used for irrigation to non-farm applications. This occurs because most urban and industrial water users can easily outbid their farmer counterparts who, given low farm gate commodity prices and generally less than efficient water application practices, can only afford to pay an infinitesimal fraction of a penny per gallon to make the business case of producing food work.

For some this is a very real moral issue, with many hundreds of thousands of homeowners wanting green lawns in desert states easily being able to outbid farmers for water. Indeed, as farmers control most irrigation systems, it can be a business decision to sell the water for more than they could earn producing food. There is nothing wrong with the prosperous lifestyle and the expectations for verdant gardens that go with it, but to seriously consider that this triggers a real water shortage is unrealistic. I suspect that the water management team of the Israeli Ministry of the Environment could only wish they could be so lucky.

On another food-water related topic, home gardening is not a particularly efficient application if there is a need to supplement natural rain. For starters, this water is processed to be potable and then is transported to homes using sophisticated and expensive infrastructure. Besides these costs, more energy is required than is normally the case involving irrigation for commercial agriculture. Secondly, for those who grow food in containers, the evaporation rate is exceptionally high, which means that water use is extreme for any benefits in food.

Although I have two barrels to collect rainwater, I use water from my well in the inefficient manner as described above. I rationalize this practice by convincing myself that many other pleasant endeavors also require energy or other resources, and this is probably less of a culprit than most. As an aside, I also must also not concern myself about all the black earth I have purchased for my garden, which originally would have been prime farm land. These two issues are raised for the consideration of those who are of the assumption that homegrown vegetables automatically have a smaller carbon and environmental footprint than their commercially grown counterparts.

To recap, the requirement for water in commercial agriculture is immense, but if it is derived from natural rain and snow as it falls directly on the land, as is most frequently the case, it is a sustainable and environmentally sound use of this resource. Similarly, when water is available for irrigation from a project that is designed for flood control, power generation, and recreation, this too should leave little room for concern. However, other applications of water for food production can be problematic. What needs to be understood by the critics of modern agriculture is that the sustainability and environmental aspects of water usage can be complex, and a simple calculation on how much water to produce a pound of beef or a bushel of wheat is quite meaningless unless the source is identified and the consequences of using it are fully understood.

17.

Animal Protein: Too Much of a Good Thing?

I feel like a bit of a hypocrite writing this chapter on the all-too-frequent excessive consumption of meat and other animal proteins, because I was raised on a farm that at various times produced a lot of lamb, beef, fresh cream, pork, eggs, and poultry. Furthermore, these same products, plus fish and seafood, are a regular but hopefully not too excessive part of my diet today. Thus I am not part of any initiative to eliminate such delicious and nutritional foods from our diets, but rather I wish to encourage consumption at sensible levels tempered by an understanding of the downsides when moderation in eating habits is not observed.

Animal proteins are a natural and perfectly healthy part of the human diet, but unfortunately prosperous people in most societies are seemingly on a never-ending path of increased meat, poultry, and dairy intake that is neither healthy nor a prudent use of finite resources necessary to feed the world. To illustrate the point on increasing consumption, I refer to the following table covering the period 1950–2007. During this period meat consumption in the United States has increased from 144 pounds per person to 222 pounds—and this does not include other animal proteins such as dairy products, eggs, and fish.

***U.S. Per Capita Meat Consumption* 1950 – 2007**

Year	*Chicken*	*Turkey*	*Veal*	*Lamb*	*Beef^*	*Pork*	*Total*
	(retail cut equiv./ lb. per person)						
1950	**21**	**3**	**7**	**4**	**44**	**65**	**144**
1955	**21**	**4**	**9**	**4**	**56**	**62**	**156**
1960	**28**	**5**	**5**	**4**	**59**	**59**	**161**
1965	**33**	**6**	**4**	**3**	**70**	**52**	**169**
1970	**40**	**6**	**2**	**3**	**82**	**55**	**189**
1975	**39**	**7**	**3**	**2**	**85**	**43**	**178**
1980	**47**	**8**	**2**	**1**	**75**	**57**	**190**
1985	**52**	**9**	**2**	**1**	**77**	**51**	**194**
1990	**61**	**14**	**1**	**1**	**66**	**49**	**193**
1995	**69**	**14**	**1**	**1**	**65**	**51**	**202**
2000	**77**	**14**	**1**	**1**	**67**	**51**	**211**
2001	**77**	**14**	**1**	**1**	**65**	**50**	**208**
2002	**81**	**14**	**1**	**1**	**67**	**51**	**215**
2003	**82**	**14**	**1**	**1**	**64**	**52**	**213**
2004	**85**	**13**	**0**	**1**	**65**	**51**	**216**
*2005**	**86**	**17**	**1**	**1**	**65**	**50**	**219**
*2006**	**87**	**17**	**1**	**1**	**65**	**49**	**220**
*2007**	**87**	**17**	**1**	**1**	**66**	**51**	**222**

***Data for 2005 are estimates; data for 2006-2007 are projections. (Source USDA)**

^Excluding veal.

On the issue of meat consumption, there is widespread agreement in medical and nutritionist circles, backed by many peer-reviewed articles, that these high levels of animal protein consumption are statistically related to increased cardiovascular diseases, diabetes, some cancers, and obesity. Secondly, as is covered elsewhere in this book, meat production other than from natural grazing and forage crops require large quantities of grain, which if the trend to increases in animal protein consumption continues, will exacerbate food shortages for many of the less privileged. Globally there is a major dietary issue looming ever larger that contributes significantly to food insecurity for the world's poor while at the same time harming the health of the more prosperous.

The issue is by no means a concern only for the United States, but most middle and high per capita income countries exhibit similar trends and health issues. China, an emerging and increasingly prosperous country, is no different. On a per capita basis, they are still far behind the United States, but their consumption of beef, pork, and broiler meat increased by 20 percent, from 45.5 pounds in 2002 to 54.7 pounds in 2007, only five years.[106] The reality of the situation is that although their average consumption levels currently are still within acceptable levels, as their incomes continue to increase, this prosperity will undoubtedly move animal protein intake to much higher levels, and with a population of some 1.3 billion, the quantities involved will be staggering. However, to single out China for going the way that so many other societies have gone is inappropriate, because this is an issue directly related to prosperity and knows no political boundaries.

The animal protein consumption issue captures a concern that I have that extends beyond food to all sustainably challenged resources as a result of the fact that the global economy is on a perpetual expansionist trajectory. In this regard I question the widespread premise that constant economic growth is the necessary and fundamental cornerstone of every functioning economy on earth. Take the United States as an example, where GDP in 1955 in constant 2005 dollars stood at 2,500.3 billion, while in 2005 it was 12,633.8 billion, a fivefold increase representing a respectable compound growth of 3.25 percent per year. This seems about right for the economy, but what about sustainability and actual improvements to quality of life?

Developed economies everywhere seem to aim at consistent growth around this level, and if not achieved, we have what is perceived to be a recession or worse. I maintain that growth is not the real issue but instability caused by such incidents as the recent housing bubble, which causes an uncomfortable disruption to the economy. Here growth is fine to get back to where things were before the bubble burst. Beyond this, why do economies such as those in Western Europe, Canada, and the United States really need to grow? There is a good argument that lesser developed economies still benefit from and require growth; thus perpetual economic expansion seems to be only a measurement of a perceived healthy economy with little or no thought given to the reality of sustainability and what the increased prosperity actually does for humanity. There seems to be modest (if any) concern that the resulting levels of consumption of items such as meat has a negative impact on society in more ways than one. I appreciate that this topic may be a digression from food, but excessive animal protein consumption might just be the canary in the coal mine as a downside to the never-ending economic growth model developed to which countries seem to be addicted.

To place the economy of today in the context of what is delivered to society, I invite readers to consider the achievements and dynamic of the US economy in the mid 1950s, which had one-fifth the GDP and a third the per capita GNP of five decades later. A half century ago, America had emerged from an incredibly expensive war on two fronts, and it had moved on to be the major contributor in the rapid reconstruction of Europe and Japan. It also built the interstate highway system, faced huge financial and technological challenges related to the Cold War, and geared up to put a person on the moon. Not too shabby for a country with 20 percent of the economic might when compared to today.

Moving ahead into the future at a continuation of the 3.25 percent annual growth rate of the last half century, which seems the level required for the economic model in play today, the US economy will be five times larger than it currently is by around the middle of this century, or a staggering twenty-five times larger than when the rather impressive achievements listed previously occurred. This means that in four or so decades, the US economy must grow at a rate of twenty-five times the actual 1950s growth in constant dollar terms. Yet this three percent GDP expansion in the 1950s was considered to be perfectly acceptable and was a reflection of a robust post war economy. In yet other words, in a few decades the US economy will be expected to expand each year by a whopping 75 percent of the entire annual GDP of a century earlier. Not only is sustainability associated with such growth a very doubtful prospect, but even if achievable, such enhanced prosperity does not likely have a parallel positive impact on the well-being of the population and could indeed have negative consequences.

Thus excessive animal protein consumption is a prime example of the unintended consequences of constant economic growth and what über prosperity brings. I hasten to add that a dynamic economy rewarding risk and entrepreneurship is essential, but the perpetual push on consumption to make it happen is unsustainable and can also have a negative impact on population well-being. An economic model based on sustainability and quality of life instead of the one that has been adopted globally (with unrelenting economic growth as its core feature) seems to be a better option.

Returning to the topic of animal protein consumption, I see the issue is one of controlling consumption or demand rather than attempting to stifle production, as suggested by many animal rights folks and others. In a free market economy, for a commodity such as animal proteins, it is demand that sets the level of supply more than the other way around. In other words, to reduce the demand for meat to acceptable levels, convince consumers of the wisdom to eat sensibly, and the demand will follow—but as long as consumers have the money and inclination to consume more, farmers will find a way to meet this demand.

For vegetarians and those who are opposed to any meat or animal protein consumption, my suggestion is that such advocates provide society with a good news story and positive overview of such a lifestyle, including health benefits. Condemning the production of these products is a tactic that mostly falls on deaf ears and clearly does not have any meaningful impact at reducing consumption. Moderation and the benefits thereof should be a more persuasive message of the medical profession, dieticians, and the concerned public.

Producers and processors of animal protein will need to develop a new mindset by facing the reality of the situation and accepting that maximizing the market share of the overall caloric intake of the population is not an acceptable industry goal. However, as producers of a premium food product, maintaining dollar market share may be more of an achievable objective, and it is the favorable bottom line, after all, which is the cornerstone of any business. Adopting such a fundamental new strategy is a huge step, I know, and probably has those readers who are part of the animal protein industry rolling their eyes. However, in reality, revenue and net income are the real goals, and ever increasing quantities at very slim margins, which often exist for this industry, do not serve any business case particularly well.

The fast food industry, with their inexpensive hamburgers and calorie-laden accoutrements can also adapt because they are seeking the largest possible share of the consumers' food dollar and not necessarily providing a maximum amount of ground meat and calories. Although a good business model until now has been to achieve market share by providing such fare, they will adapt under sufficient consumer pressure and regulation. As can readily be observed over the past few years, salads and other somewhat healthy offerings are quite commonplace in such establishments.

On the regulatory front, the calorie content and other facts regarding nutrition should increasingly be made a prominent part of the packaging and posted appropriately in the eating establishment. Instead of spending energy rallying against the meat production and fast food industries, it would seem more productive for folks to demand an accelerated movement toward healthy offerings in fast food outlets, and voters should encourage elected officials to respond with supporting legislation and regulation.

There is perhaps a parallel in the alcohol industry. In recent decades the alcohol producers, the concerned public, and the government came closer together and worked on a common message of moderation, and although the serious issue of excessive alcohol consumption is still with society today, at least there is a somewhat consistent message on moderation, which makes the likes of Alcoholics Anonymous particularly effective. All parties support their fine work, and there is no distortion of their message, which is to accept that alcohol abuse exists but can be mitigated through practical approaches at the level of the individual. Polarity on issues by critics of the food and beverage manufacturing sector and the industry itself that is in the defense of the status quo, leads nowhere. Meaningful dialogue and an effort to seek common ground are much more productive and provide an opportunity to achieve positive change.

18.

Animal and Poultry Welfare

Humans are omnivores and have a digestive tract with many characteristics, starting with teeth that are well adapted to maximize the nutritional benefits of animal proteins. This dietary flexibility has been with humans from the time of early hunter-gatherers, but today it comes with the troublesome fact that something with a life that humans can identify with must first be confined and then perish so folks can have a bit of meat on their fork.

Furthermore, the end of such life is often abrupt and is by no means pretty. Understandably, this dynamic gives rise to discomfort by thoughtful individuals, but on the other hand such an ingrained human trait is not going to disappear, and thus the consumption of meat and indeed eggs and milk, which are also associated with animal welfare, are a fact of life. Michael Pollan does an excellent job of capturing this collision of natural human dietary customs with the socially conscious side of our advanced reasoning in his book *The Omnivore's Dilemma*. I will attempt to take the middle ground on these very divergent positions—and perhaps end up with both animal rights advocates and the meat and poultry industry in disagreement with me.

Starting with the hunters of the hunter-gatherer epoch, I suspect that this dilemma was not of great concern because a meal was a meal, and something had to be killed to ward off hunger for the individual and the community. The option to pass on meat in favor of a vegetarian diet probably did not exist. Also, the thrill of the hunt was a reward in itself that shoppers of today have little chance to experience. However, the chasing, trapping, and killing would have probably have been much more brutal than the organized slaughter of animals today. For example, little regard would have been given to considering nursing mothers and their offspring. Also, with primitive weapons, the actual act of killing could not have been quick and without terror and pain, particularly for larger mammals. Though I am not going to portray the killing floors of slaughterhouses as a pleasant place, they are a vast improvement over the way that our hunter forefathers probably ended the life of their prey.

Over the years as an agriculturalist, I have had the opportunity to observe the slaughter of hogs, cattle, and sheep in industrial scale slaughterhouses. In New Zealand the spring lamb slaughter, butchering and cutting up the carcass, was at a rate of one a minute, for three shifts per day and seven days a week. In all visits, efforts were made to keep the animals as calm as possible until just before the slaughter. The animal was in a holding pen with fellow beasts, where there was some comfort in numbers, and then a narrow passageway led them to where they were stunned unconscious and then slaughtered. From an existence that brought some stress to be sure, to the possible realization of impending doom would have been a matter of seconds.

The steady slaughter of animals is far from a pretty sight, but it is far more humane than many writers make it out to be. One fundamental factor is that calm animals are much easier to handle, are more predictable, and are less likely to injure themselves than if they are in a frantic panicked state. Also, there is a school of thought that an adrenaline-charged carcass from a frightened beast will suffer some quality loss because of the tension in the muscles, which toughens the steak or roast. Thus a standard slaughterhouse protocol is to maintain the best possible environment for the animals as they proceed to their death. By happy coincidence, the somewhat humane treatment of the animals is also best for the bottom line.

Yes, millions of animals in the prime of their lives are systematically slaughtered to provide food for humans. For those who find this intolerable, I can only suggest that they let their conscience be their guide and, because they have ample choice, choose a vegetarian diet. However, to campaign against the entire animal protein industry is an unfair imposition of individual values on the general population that can accept that meat products are a natural part of the human diet and that death is an unavoidable element of the process bringing this type of food to the fork. Furthermore, the quantity of food entering the global system would be significantly decreased if the grasslands of North and South America, New Zealand, Australia, and elsewhere, that have little utility for most types of agriculture, could not be utilized for meat production.

Another topic of real concern and debate by animal welfare advocates is the transport of animals from ranch, farm, or feedlot to the slaughterhouse. Generally most humans, including farmers, truck drivers, and slaughterhouse workers, do not like to experience animals suffering unnecessarily. It also should not be overlooked that animals are valuable property, and unless they are delivered to market in top condition, there is a substantial loss along the way to the farmer, trucking company, or the receiving plant.

As an observation, once off the farm, every animal is for some time the responsibility of a truck driver in much the same way as new cars are in the possession of drivers when moving the product from factory to dealer. If such a driver arrived at his destination with dented cars with any regularity, this would be grounds for dismissal. Similarly a driver of a livestock transporter who had a history of frequent injuries on delivery of his cargo would face the same consequences. In my days on the farm, we had a few hundred beef animals delivered each year as young animals, to be later transported to market. I cannot recall ever seeing an animal that was delivered with a broken leg or serious injury, nor do I remember any issues of such incidents when transported from the farm to market. The drivers were a professional lot who moved livestock for a living and seemed to know what they were doing and how best to handle their living cargo.

However, it is quite apparent that an animal weighing a thousand pounds and having a high center of gravity with long legs (which are originally suited for fleeing ancient predators and moving over large distances on sparse grasslands) is not in a good place when being transported in a vehicle with a driver that might need to brake unexpectedly or take sudden evasive action. There are most definitely injuries and some loss of life. However, those behind the wheel understand this and can do a lot to prevent unnecessary harm to the animal. I therefore do not share the opinion of those who are critical of the meat industry, stating that the maltreatment of animals and poultry in transport or at the slaughterhouse is a systemic and widespread problem aggravated by uncaring workers and their managers.

One area of animal welfare that rightfully receives substantial attention is the rearing of poultry for both egg and meat production. There seems to be a belief that chickens somehow have lifestyle aspirations like humans and consider their caged existence cruel and unusual punishment. Though I fully agree that intensive commercial broiler and egg production is very far from being hen heaven, perhaps it is not a life in poultry purgatory, either.

Let us explore the entire history of the chicken, which can be covered in a couple sentences. About ten thousand years ago, the junglefowl[107] (it looks somewhat like a pheasant) probably realized that if they hung around humans, they were fed, watered, and protected from the elements and predators. When receiving these benefits, they became property and gave up some or all of their freedom to roam at will. While this arrangement may have seemed too good to be true for the bird that evolved into the domestic chicken, their side of this rather one-sided bargain never was an issue, except for the last few seconds of their existence. All in all for the living flock, it was not a bad arrangement.

But what about the loss of freedom, which has probably become a bit harder to take as farming evolved into the large and efficient poultry operations that exist today? But really, how much worse is it for the poor bird? In the good old days of heritage breeds on small farms, guess where chickens were housed? In a chicken coop, which, by its very name, suggests confinement in a small space. Was this a cost-cutting measure or a good way to raise chickens? I maintain mostly the latter.

On this I have personal experience from my childhood on the farm, where we had thee largish chicken houses for hundreds of laying hens that had the opportunity to freely go outdoors whenever they wanted to in the summer. However, if the weather was inclement, if there was any threat from predators, or when night fell, they all gravitated to their respective buildings where hundreds preferred the close confinement. When indoors after dusk, the hens chose to roost tightly packed together on raised platforms and thus leaving the floor uninhabited. Indeed this natural tendency to crowd together was sometimes a problem to the point that some perished through smothering at the bottom of a mound of birds. This was not an uncommon occurrence, particularly when large numbers of younger birds were given complete freedom of movement and for some reason panicked.

The point is that chickens seem quite comfortable being close to each other, and confinement is not the issue some might think it is. Humans accept small spaces comparable to a chicken in a cage more often than they might realize. Let us assume that a person weighs about fifty times as much as a chicken. If we multiply the typical room a caged chicken has by this amount, this is comparable to the space that humans allocate themselves in the classroom, during sporting events, in transportation (be it plane, car, or bus), at the movies, at live performances, and in bed. Even an office cubicle is proportionally not much larger.

Then there is the issue of cannibalism in chickens. Back on my childhood farm, although the birds had free range whenever they wanted it, they somehow were prone to pick at each other; presumably this was the establishment of a pecking order. Once blood was drawn, which may not have been the original intent, the exposed injury appeared to be an open invitation for the entire flock to get into the act, with unhappy results for the poor bird that clearly was now at the bottom of the pecking order in more ways than one.

To counter this behavior, a bright red hot pepper paste was applied near the wound of the affected chicken, and generally the aggressive behavior stopped. (In the useless information category, this would seem to indicate that chickens have a modicum of intelligence, have refined taste buds, and are not color blind.) The point I am trying to make is that this cannibalistic trait is best controlled if birds are alone or in very small groups, which is an advantage of caged confinement.

Another positive coincidence of positive animal welfare and farmer prosperity is the very real link between production and a non stressed animal. There is some wisdom behind the 1907 Carnation Milk slogan, "Milk from Contented Cows." There is ample research that the effort to provide excellent feed and an appropriate environment results in more milk per cow, and some even maintain better milk. Though the confines of a dairy barn may not seem like a particularly pleasant place to the casual observer, it seems as if the cows do not mind it with their ample diet, constant access to water, and protection from the elements.

Perhaps if cattle could rationalize their situation, they would hark back to their more primitive bison ancestors and conjure up a worst-case ancestral scenario, such as being surrounded by wolves in a January blizzard with three feet of snow covering the dried remnants of the previous summer's grass and bitterly cold temperatures. Again, confinement may not be such an unpleasant ordeal for all types of domesticated animals compared to the less positive aspects of life in a natural setting. Even today, farmers would be charged with animal cruelty if they subjected their domestic animals to the same conditions that countless deer face whenever there is a harsh winter.

Similarly for egg production, every profitable poultry farm achieves close to an egg a day per laying hen when in their prime. This is quite an amazing feat that goes on for several months and is somewhat parallel to a mammal creating an offspring every twenty-four hours. But if life was really unacceptable for the hen, egg production would plummet—a situation which would not be good for the farmer. Back on the farm where I was raised, every effort was made to avoid situations that would cause the birds to panic or come under any stress. When young cousins were visiting, the common refrain was, "Don't scare the chickens or they will stop laying eggs." Thus even the animal rights groups should take some comfort in the fact that chicken farm profitability is directly linked to at least reasonably contented birds. The same principle applies for meat-producing animals, be they pigs or cattle: adequately provide the basic necessities of life, and they thrive without any signs of stress.

Incidentally, for some, there might be reassurance when purchasing eggs or chicken to read on the packaging that it is "free range," which might conjure up images of a life of freedom to roam outdoors at will and live partially off the land. The USDA definition, which permits this declaration, reads as follows.[108]

> **Free Range or Free Roaming:** *Producers must demonstrate to the Agency that the poultry has been allowed access to the outside.*

Poultry producers are in the business to make money and probably have many thousand birds, so it is inconceivable that access to the outside comes close to providing the extensive space so that all chickens can enjoy the great outdoors with total freedom to wander as they please. Most likely there will be access to a confined, open-air section that enables a small proportion of the flock to go to at any one time.

I do not pretend that in an industry as large and diverse as the raising, transporting, and slaughtering of the nation's cattle, pigs, sheep, and poultry, there will be incidences of poor management practices that lead to the unnecessary suffering of the animals or birds. However, while inexcusable, such operations cannot be expected to be profitable or remain in business unless they adopt better animal husbandry practices which will have a positive impact on the bottom line. These examples of abuse and mismanagement understandably garner the attention of animal rights groups and unfortunately tarnish the image of the majority of farmers who operate very much with the welfare of their animals or poultry in mind, if for no other reason than to enhance profits.

19.

Hormones and Antibiotics in Livestock Production

There is much controversy regarding antibiotics for both four-legged livestock and poultry, as well as growth hormones for beef and dairy cattle.

Regarding antibiotics, there are two types of application, the first being administering a sick animal. This is no different than attending to the immediate health needs of an ailing person and is necessary to mitigate suffering and perhaps save the life of the animal. There should be no opposition to this practice, and if there is, it is without any basis. It should be noted that in both the United States and Canada, antibiotics in milk are prohibited, and to ensure their absence, a testing protocol is in place. If an ailing milk cow requires an antibiotic, she is isolated from the rest of the herd until her milk is tested and determined to be free of this substance.

Where there is an understandable controversy is the practice of the preventative application of antibiotics in the daily ration of livestock and poultry. Here the issue does not particularly relate to the direct health implication of the food but rather the perpetual opportunity for wayward bacteria to mutate and develop a resistance to such antibiotics, which are often identical to ones used in human medicine. Furthermore, even for bacteria normally only found in livestock and poultry, resistance to antibodies gives rise to the possibility that if a trans-species transfer to the human population occurs, there may be a lost opportunity for an effective defense.

Epidemiologists and related science professionals are almost universally against this practice for the reason outlined above, and they have evidence to back this position. Although I tend to support many advanced and sometimes controversial agricultural practices, the constant use of antibiotics is one where

I see the risk is much greater than the benefit. It also may not be any real advantage to producers.

The largest exporter of pork meat in the world, Denmark, has had a ban on the non therapeutic use of antibiotics in hog production since 1995.[109] According to most accounts, the industry is still thriving and reports a much lower incidence of antibiotic-resistant pathogens in pork compared to other European nations that have no such restriction in effect. The decision by the Danish government to implement this regulation was particularly courageous because pork accounts for 6 percent of the value of total exports from that country.

The regulation was possible because of the existing high standards of the industry, with excellent farmers backed by state-of-the-art veterinarian support, thus making the elimination of antibiotics in the daily hog ration possible with little if any negative impact on the economics of raising pork. Given that the North American industry is equally sophisticated, with the added advantage of a decade and a half of Danish experience to draw upon, there seems to be no reason why a similar regulation on this side of the Atlantic would not also work.

Another controversy is the use of hormones to enhance the productivity of livestock. Perhaps the most controversial application is that involving dairy cows. I will not get into the science behind consumer concerns regarding this practice, as the research on this issue is extensive and complex but inconclusive. In this regard the Program on Breast Cancer and Environmental Risk Factors in New York State[110] has prepared an excellent and balanced overview of health issues relating to this procedure in dairy barns.

Stepping back from the human health issue, it is interesting to note that this practice is banned in the European Union, Canada, Japan, Australia, and New Zealand. Furthermore, an estimated 10–30 percent of US dairy cows are treated with hormones. Considering the industry seems to thrive without this practice, some evidence exists that these hormones could be harmful to consumers, and there certainly is a lack of consumer confidence in such milk, an argument can be made for an outright US ban as well.

The growth-enhancing practice of hormone treatment of beef is somewhat more difficult to dismiss because there seems to be very little peer-reviewed research demonstrating any dangers to human health. Although some informed consumers are against the practice, there are demonstrated cost benefits to the producer in terms of weight gain per day. Unlike hormone use in dairy herds, with the United States as the only developed country permitting the practice and with modest industry uptake, the use in beef cattle is only banned in the European Union. Thus the practice is much more widespread globally than just the US milk industry. Without reliable scientific work demonstrating health risks and some plausible research suggesting there should be concern, it is doubtful if this practice will be banned anytime soon. For readers with concerns, organic meat products are at least certified to be hormone free.

As far as pork and poultry products are concerned there is little or no economic benefit to the use of hormones in these industries and thus no interest on the part of the farming community to introduce the practice.

The USDA website[111] on food labeling states the following.

> *Hormones are not allowed in raising hogs or poultry. Therefore, the claim "no hormones added" cannot be used on the labels of pork or poultry unless it is followed by a statement that says "Federal regulations prohibit the use of hormones."*

Hopefully this demystifies some of the basic issues regarding the use of antibiotics and hormones in livestock production. The subject is highly controversial and will remain so; it is of real concern and interest to consumers who might find this preliminary overview useful should they attempt to delve more deeply into the subject. A quick check of material on the Internet will provide a wealth of information, some well researched and others more of an emotional nature, covering both sides of the issue.

20.

The Pros and Cons of Backyard Egg Production

There is a growing interest in transferring a small part of agriculture into an urban setting with backyard egg production. Much like gardening, this reduces the distance from farm to fork to a matter of feet.

My farmer instincts were thus in favor of having a dozen or so hens supplying fresh eggs every day just outside my door. As I live on a large property quite away from neighbors I decided that it was possible to take up the challenge in spite of the fact that there is a bylaw prohibiting such a practice. Before I made the commitment, I decided to research the concept although it seemed so promising at the outset.

First there was the requirement for suitable shelter combined with a secure outdoor enclosure if there was to be a modicum of freedom for the birds. In this regard, prefabricated hen houses for small flocks could be ordered from internet suppliers but at a cost of several hundred dollars. Building one's own was the next option, and the price came down to half. I then noticed that these were single-boarded, warm weather structures only, and if I wished to keep my chickens over the winter, insulation and heating must be provided. This added substantially to the original cost and meant added power bills (and a carbon footprint) in the winter.

More research brought home the realization that chicken feed is a mouse, squirrel, and rat magnet that could be mitigated if the chicken coop was properly designed to keep these rodents from gnawing their way into the structure. Vermin control also meant that the he outdoor run had to be sturdier. This further contributed to mounting up-front expense.

Then there was the feed which was readily available at the nearest farm supply store but was some miles away. As one might expect, they charged a substantial premium over what a commercial chicken farmer ordering the feed by the ton would pay. The same inflated price applied for oyster shell, which a productive chicken needs to provide the calcium for the egg shells. Also because of the vermin issue, the feed also had to be stored in rodent-proof containers.

Also, there was the task of at least once daily feeding and watering of the birds, gathering the eggs (the fun part), and a weekly cleaning of the coop. After the first year, there would be added cost and time for maintaining the structure and enclosure. I travel a fair amount, and thought the pets in the house demand attention if not taken care of, chickens out of sight, particularly in the winter, could end up being inadvertently neglected by others in the family who perhaps did not have the same enthusiasm for the project as I would have. This could be an issue with not only bird welfare implications, but it may also lead to some disharmony with the family.

While the city would consider these birds as livestock, I knew that they would soon acquire names and become quasi pets. Laying hens have a productive shelf life of about a year, and then egg production drops substantially as they go into a lengthy retirement that can last a couple years or longer. Keeping in mind that the entire exercise was to keep the family in fresh eggs, the reality of productivity versus a retirement home for chickens presented a real dilemma. Was I to play the role of farmer and have no qualms of retiring the birds as our Sunday dinner? This was obviously out of the question, because particularly the children could not bring themselves to dine on dear old Henny Penny. Playing the role of backyard executioner was equally distasteful, and besides, there was a city ordinance that all household animals, when the time came, had to be put to sleep by a qualified veterinarian. This took some of the emotional sting out of the equation, but research came up with a cost of thirty-five dollars per bird for euthanasia at the local veterinary clinic. Getting them there without the proper crates could also be a challenge.

In spite of all this, it was still a doable endeavor if money was no object and if I chose to ignore the carbon footprint of heating a small building for a tiny flock while a poultry farm with close confinement mostly relies on the birds alone to heat the shelter. A second issues relating to the carbon footprint was that I would be transporting the small quantity of feed fairly long distances in a ton and a half vehicle, which would involve a lot more energy per bird that for the commercial farmer who had feed delivered in large trucks that weighed less than the load in the back.

But money *was* an object, and I did a few "back of an envelope" calculations on how things might work. The best a hen will do on a well-managed farm is about 200 eggs a year given that time it takes for a chicken to mature and the fact that at one year of age peak performance is past and the bird is sent off to market. As I anticipated a lesser result, I calculated my realistic yield would be closer to 150 eggs per bird. Twelve birds would produce 150 dozen eggs per year, or about enough to meet our family needs. Less fresh but perfectly acceptable eggs are available at around two dollars a dozen, which means my annual harvest would be worth about three hundred dollars.

Expenses, starting with the veterinarian euthanizing option alone should I go that route, would have been $420. Then with the purchase of the baby chicks, feed, oyster shell, electricity for heat and light, initial cost of the hen house spread over five years, and maintenance, I calculated that the cost of my eggs would run about three times the store-bought version. In addition, while the thought of chickens clucking only steps from my door in summer brought back happy memories of life on the farm, I also realized that in the cold of winter, things might not be quite so pleasant, plus the effort to get the children to properly attend to the flock in my absence when traveling reality entered the equation. Then there was the dilemma that either I let the hens live out their lives in ever declining productivity or facilitate their execution when their productivity lagged. Much to the relief of my wife, I decided that common sense should prevail and that I would abandon this farming fantasy.

For readers who live in a jurisdiction with bylaws permitting back yard flocks of chickens and are contemplating such an initiative as I did, I strongly recommend that they first go through a similar reality check. If after this the benefits of fresh eggs plus the joys of having your very own mini farm outweigh the costs and inconveniences, then by all means go for it, and good luck!

21.

Food Expectations of North Americans in a Hungry World

I suspect that for many readers the concept of farm to fork will not automatically include a vision of feeding the world; this is an opportunity and challenge facing the same farming industry that readers of this book rely on to provide the bulk of their nourishment.

The global population passed 6.8 billion in early 2011. Expectations are that it will exceed the seven billion mark within the next two years and quite possibly climb to nine billion by the middle of this century.[112] The farmers of this world are facing a daunting task to feed so many people.

Such population numbers are quite difficult to relate to in a meaningful way, but I will attempt to illustrate how much food is required to feed all these folks each day. According to the Oklahoma Wheat Commission, there are about a million kernels in a sixty-pound bushel of wheat. To provide each person living on this earth with a single grain of wheat, it would require 6,800 bushels weighing about 200 tons. This represents an ample load for nearly seven large transport trucks of the kind one regularly passes on the highway. Just think about it—seven big truckloads of wheat would provide a minute quantity of food for each person on this earth, barely sufficient to get stuck in one's teeth!

Now what would be required to provide every man, woman, and child with enough calories for a sustainable existence, assuming an average daily requirement of two thousand calories which, is about what two pounds of wheat will yield? For illustrative purposes, this means that every day the world's basic food needs to feed nearly seven billion people is the equivalent of fourteen billion pounds of wheat, or seven million tons. Again assuming thirty tons per truckload, this is equivalent to 225 thousand large loads are required each day to feed the global population in 2011.

This is a meaningless number unless we can visualize such quantities. Take a typical novel, with about four hundred words per page. This means that in a six-hundred-page hardcover book, there are close to as many words as truckloads of food required each day to feed the world. Dan Brown's bestseller *The Lost Symbol* has less words.

One more way to visualize so many large trucks would be to consider if one is traveling down an interstate and meets ten trucks per minute, which would be considered heavy transport traffic at six hundred trucks per hour. Passing the 225,000 trucks would take about 375 hours of driving and 22,000 miles. For those who drive that many miles a year, the approximate total of all the trucks met during this time, if loaded, would feed the world for only a single day.

I present the big picture here because most consumers of food quite understandably look to agriculture—and the subsequent transport, storage, processing, and marketing—as being there only to meet their own personal food needs, those in their community, and perhaps very little further afield. Within this somewhat limited sphere of nutritional needs, consumers have rather exacting opinions on what is appropriate regarding the quality, freshness, source, means of production, and other attributes of food. This is perfectly understandable, and for the typical consumer there is little need to take the global situation into consideration when acquiring one's food. However, should consumers have issues with large-scale commercial agriculture or advanced farming practices, which perhaps appear to do little for them, it is important that there is an awareness of the immense quantities of food that are required to feed the world, and this can only be achieved with an efficient and advanced agricultural sector that, by necessity, often mass produces food to take advantage of economies of scale.

Returning to the practical aspects of meeting global food security, the temporary wheat shortage of 2007 was because of a poor Australian crop, and the ripple effect hit other major food crops such as corn and rice. Though there were no actual food shortages, just the concern about the prospect of such a situation not only set off food riots, but it encouraged hoarding, which made a somewhat worrisome situation very much worse.

Meeting global food requirements is a precarious state of affairs, with the difference between surplus and perceived scarcity being only one bad harvest in a single major food-producing country. Thus those who rally against the way farmers everywhere go about their business to somehow provide sufficient food to feed the world should think carefully about getting what they wish for when they advocate the elimination of genetically modified crops or other advanced agriculture technologies.

22.

Fish, Fishing, and Fish Farming

My commercial fishing career is limited to the occasion when I volunteered to spend a morning alongside full-time fishermen on a mackerel run a few miles off the east coast of Nova Scotia. While very much a novice, I am rather proud of my record haul that day, which was fourteen one-pound-plus mackerel on a line, with twenty hooks that had been in the water for only a matter of minutes. In spite of this achievement, I am not particularly well qualified to write about the finer details of fishing when compared to countless others who are much more familiar with this topic. However, as this book is intended to provide an all-encompassing overview on the topic of food this chapter has been included.

Arguably, a freshly caught wild fish such as salmon is as close as one can get to an ideal animal protein food. It is not only rich in nutrients such as omega-3 fatty acids but is also an exceptionally delicious and easy to prepare meal. For those who have ready access to wild-caught fish and can afford to pay the extra over the normally lower-cost, farm-raised variety, the choice is clear.

Unfortunately the wild fish in oceans, lakes, and streams are in many cases, going the way of the North American bison and countless other animal and bird species, which are either extinct or reduced to a fraction of their original range and population thanks to a losing battle for land resources with humans. Though not competing with humans for space, fish face unremitting pressure of overharvest, particularly in the oceans that are mostly beyond any meaningful jurisdictional oversight and conservation enforcement. The unfortunate fact is that the global fishing fleet, with large-scale vessels and advanced technology, regularly harvest the oceans at a faster rate than fish stocks can replenish. For the individual fisherman, this makes perfect economic sense. The resource is there for the taking, and any efforts regarding conservation or limiting their catch will simply mean that the holds of the ship remain less full while uncaught fish remain in the sea and can be captured by the next fishing vessel that comes along.

Even beyond individual fishermen, it is in the short-term national interest of many countries to either encourage or turn a blind eye to the goings-on of their fleets on the high seas. If such advanced and normally environmentally savvy countries as Japan and Norway still allow whaling, the oceans seems to be clearly up for grabs for whoever comes along. Whales are surface mammals that are readily observed with recognizable family structures and a formal means of communications, and thus they are perhaps mankind's favorite resident of the high seas. If such a species faces the fate of a harpoon, what chance do invisible, less well understood, and often not very pretty or interesting fish populations have?

Unfortunately once fish populations have become depleted, their recovery is exceptionally slow, and they may never reach their previous levels, particularly if fishermen return to international fishing grounds as soon as there is a glimmer of hope that the stock is naturally replenishing itself. A prime example of this is the 1992 collapse of the cod fishery of Atlantic Canada.[113] For centuries, this region was one of the richest fishing resources in the world and not only included cod but a host of other commercial species as well. Although the fishery was mostly within Canada's two-hundred-mile territorial limit, which legally gave the country complete control over the resource, the population of cod and other species were so severely overfished that the government was forced to completely close the fishery and put forty thousand people in the industry permanently out of work in Newfoundland alone. Now nearly two decades later, this fish population remains in a delicate state and is much too fragile for the reintroduction of commercial fishing of any consequence.

If fish are to remain as an important and healthy part of our diet, the alternative is fish farming. Though there are many against this practice for sound environmental reasons (which is a variation of the same theme are as is the case with farming), perhaps there is room for some compromise. If the wild fish catch was expected to remain large and sustainable, I would probably side with the environmentalists and question the need for the artificial rearing of fish. Unfortunately, as outlined earlier in this chapter, the perpetual bounty of the seas is a very remote possibility. Therefore, I see no option but to embrace best practices and encourage a robust fish farming industry for any species that adapts to this technology.

Failure to do this will drive the price of the dwindling stocks of wild fish to levels unheard of today, to satisfy the prosperous who can afford to maintain this element of their diet. This will make it economic for commercial fishermen to still make a living even though the catch is poor—and while completely depleting challenged fish stocks. Besides feeding humanity, fish farming, through lower prices and basic availability, contributes substantially to saving viable wild populations for future generations to harvest sustainably.

For opponents of fish farming, the pollution of offshore waters near the pens, disease, parasites, and escaped fish diluting the local wild gene pool are all real concerns. To address these issues, mitigating practices are being introduced and therefore the outright condemnation of the entire industry may be inappropriate.

For example, the practice of towing large pens at very low speeds offshore eliminates the buildup of pollutants, and with the flow of water through the pens, the incidence of disease and parasites are also reduced.[114] Even for stationary pens near the shore, infrared sensors are placed beneath where feeding occurs to monitor when the fish have finished their meal, thus eliminating overfeeding and the waste and unnecessary pollution of decaying rations. For the material that does accumulate under the pens, various organisms indigenous to coastal waters have been found that flourish in this environment, and when introduced as a viable population, they greatly reduce the negative environmental impact of this waste.

Another concern associated with fish raised in captivity was the large quantity of their wild cousins, such as the anchovy, that went into their ration compared with the edible product that eventually made it to the forks of consumers. Advanced fish farms today maximize the use of agricultural protein (primarily soy meal) and chicken oil, a low-value by-product of the poultry industry to the point that freshwater species such as tilapia require less than a pound of fish in their ration per pound of gain. Marine species have not yet achieved this level of efficiency, but they have improved the ratio substantially in recent years and are continuing to make further progress.

As a consequence of declining stocks of larger, more desirable species, the number of anchovy and other small wild fish have increased in certain regions. This leads to a potentially sustainable harvest that may be desirable to keep these expanding populations in balance. Though these smaller fish make perfectly good human food, their preparation is labor intensive, and also they are considered by most to be an inferior product when compared to the larger species favored by commercial fish farms. Furthermore, as with any fish, a large proportion of an anchovy, such as the head and internal organs, will probably be discarded as inedible for humans. When used as a ration for fish farming, the whole fish is used as feed.

It should not be overlooked that wild fish do not simply "happen." It is estimated that for every pound of tuna that humans consume, about a hundred pounds of lesser fish were ingested.[115] Thus if undesirable fish flourish because of depleted wild stocks of commercially viable species, is their use for aquiculture purposes as unacceptable as opponents of fish farming maintain?

Although those against fish farming do raise real environmental issues, are these mostly for outdated practices that will become less and less of a factor as time progresses? On the other hand, the industry has much to contribute to both feed the world and to take the pressure off the overfishing of wild stocks. With this in mind, I encourage those who demonstrate zero tolerance for fish farming to revisit the substantial benefits of this activity and to promote best practices in place of advocating closure of this promising industry.

23.

Background on Farming of Selected Food Products

Wheat

By now readers may have guessed that I have a special interest and appreciation for both wheat as a food and its production. The relationship goes back to my youth on the family farm. Although we produced barley, oats, and sometimes rye, wheat was my favorite grain. Early in the growing season, as it was blanketing the field, it was the darkest and richest green of all grains, and it looked like a very lush lawn. When harvest came it was a purer gold color in the field compared to other grains. Also, the fact that it all left the farm to be consumed by humans, while most of the other grain was destined to feed our livestock, also added to its special status. I marveled at the possibility that someone across the country or world might enjoy the bounty of our fields, or that this wheat could find its way into the flour or bread that we purchased at our local grocery store.

In university my appreciation grew as wheat, along with its less versatile cousins of oats and barley, was considered the crop of last resort. This description, in agricultural terms, was to honor the grain because it flourished where climate was too cold, too hot, or too dry for corn and most other food crops. Without this robust characteristic, most of the Canadian Prairies—and indeed a lot of the US Great Plains as far south as Texas, the drier regions of Australia, and elsewhere—would produce much less food. When it comes to global food security, wheat alone can contribute in a very substantial way on land that would otherwise be beyond the climate zone of commercial food production. Indeed, those concerned with feeding the world in the United Nations estimate that 20 percent of all calories consumed by humans are wheat products.

Field of Ripe Wheat Ready for Harvesting

Furthermore, it is the most versatile of grains suitable for the pasta of Italy, the noodles of China, and the breads of an infinite array of different possibilities. In the Middle East, wheat is consumed as cuscus, bulgur, or burgol. On the Indian subcontinent it is naan or roti. All of these products are staple food to billions. As readers are well aware, wheat is in a host of bakery products, as well as some of the more nutritious of the breakfast cereals. These various products require different kinds of wheat, and when these usage requirements are combined with climatic differences, there are several hundred varieties of wheat in the United States and Canada that fall into six basic classes.

In terms of total production, the dominant wheat grown in the United States is hard red winter wheat, which is mostly produced in the Great Plains east of the Mississippi River; lesser amounts are grown in western Canada. The advantages of winter wheat is that it is planted in the early fall and partially develops before winter, when it remains dormant until spring and then matures quickly, taking advantage of early, favorable moisture conditions before the extreme heat and perhaps limited rainfall of summer. The protein content of hard red winter wheat varies, but it is generally considered to be a good milling and baking wheat suitable for bread and all-purpose flour.

Hard red spring wheat, which as its name suggests is planted in the spring, has the highest protein content of all wheat and is considered to be superior for milling and bread baking purposes. This wheat is favored in Canada and the states bordering the Prairie Provinces, where the climate can be too harsh for winter wheat. This is Canada's principle wheat export.

Soft red winter wheat is primarily grown east of the Mississippi River in the United States, as well from Ontario and further east in Canada. It has a lower protein content then other wheat classes and is used primarily for cakes, cookies, pastries, and crackers.

Durum wheat is the hardest of all wheats and is primarily used to make semolina flour, the main ingredient for pasta products. It is a spring wheat and is grown in similar conditions to hard red spring, but with somewhat reduced yields, primarily in North Dakota in the United States and Saskatchewan in Canada.

Hard white wheat is similar to the hard reds in quality and use, except for the color genes. The flour and baked products using this wheat are almost invariably of the whole wheat variety and provide all the nutritional advantages, but they appeal to consumers who prefer white bread, rolls, and other wheat-based products.

Soft white wheat has very similar properties to the soft winter counterpart as far as baking and milling qualities are concerned. It is primarily grown in the US Pacific Northwest.

As to the actual farming of wheat, and indeed of all field crops, there is a widespread belief that modern farming techniques, involving fertilizer and chemical applications are relatively recently adopted practices and the issues surrounding these are new and the consequences have not been fully studied or understood. In contrast to this perception when I was growing up on the farm several decades ago, the extensive use of chemicals and fertilizers on large fields was already very common.

Much has been learned about the optimal application of these inputs and other practices over the years from both from the environmental and agronomic points of view and in the process a lot has been accomplished to mitigate the negative impact of farming. For example, the disaster years for soil erosion were the "dust bowl" period in the Great Plains of the 1930s that have never been repeated. Indeed with modern tillage practices starting with replacing the plow with other less intrusive cultivation implements and more recently with minimized tillage, wind and water erosion are much less of an issue today than in the past.

With the advent of precision farming technology, utilizing a type of GPS,[116] [117] a sophisticated version of the units found in cars, satellite data provides a very precise soil and vegetation map of individual fields. Armed with such technology in conjunction with advanced farm equipment, the farmer applies the optimum amount of seed, fertilizer, and chemical inputs that matches the ability of any given piece of ground to utilize such inputs in the most efficient manner. This helps the farmer's bottom line and reduces the amount of these materials that is applied on both a per acre and per bushel yield basis.

One unfortunate aspect of applying no-till and GPS technology is that the equipment is generally of a larger scale and is very costly; therefore it is not a viable business option for most of the smaller farmers, and certainly for the 50 percent with a turnover from food producing activities of less than ten thousand dollars per year. This is the reality of many types of technology application in industry and elsewhere in the non farm economy. It is invariably difficult to achieve efficient levels of production through cutting-edge technology without a fairly substantial scale of operation. Economies of scale matter a lot when it comes to the production of wheat and other grains. In this regard, contrary to some common beliefs, larger farms that can adopt advanced technologies are more likely to be environmentally more benign than their smaller counterparts.

Even with these advances in technology that to some extent will mitigate the environmental impact of farming, the general concern that farming of crops such as wheat is a harmful activity to the environment and somehow must be changed. To be certain, all human activity, be it urbanization, transportation, manufacturing, or farming, have a negative impact on the environment. If the large population on this planet is to exist, all of these activities are a necessary reality that must be faced, with every effort to mitigate their impact where economically possible. In this regard, wheat farmers and others in the business of raising food from the country's fields have made some impressive technological adaptations with the promise of more to come.

For wheat the most promising breakthrough will be the introduction of genetically modified varieties. As stated previously, wheat already grows where it is colder, hotter, and drier than most other crops. With the advent of climate change, increased drought and heat tolerance, along with a better utilization of fertilizer and chemical inputs, all through genetic modification, will make this an even better crop to feed a hungry world.

Returning to my personal connection with wheat as a diplomat in Germany, China, and South Korea, wheat was an important element in our exports to these countries. This provided a great exposure to the intricacies of international trade of this commodity. Wheat even got me arrested on one occasion. Early in my first posting to Germany, where I was agricultural attaché, I spent a couple months living with a family in Kassel to pick up the lingo. I was staying with the chief forester for the area, and their house was somewhat rural with a wheat field across the road. It was June and the kernels were in the "dough" stage (when removed from the head, they were quite edible as it is basically in the same state of maturity as the sweet corn that we eat on the cob). As I was enjoying memories of a similar snack from my boyhood, I was suddenly facing two border police who were ubiquitous in that region at the time because the then highly sensitive East German border was only a few miles away. I was unexpectedly in some difficulty, with no personal identification in my possession (a serious issue at that time and place), no ability to effectively communicate in German, and being apprehended as what appeared to be a desperate vagrant feeding himself close to the border where unwanted refugees were not unknown.

Fortunately the forester's house was nearby, and when I pointed to it and mentioned his name, it resonated with the police. They took me to the door, where identification was quickly made. After presenting my passport, my status changed from suspected fugitive to eccentric Canadian, and my troubles quickly evaporated.

I also had a stint as director general with Agriculture Canada, head of the Grains and Oilseeds Marketing Bureau, which among other responsibilities was the interface agency between government and the Canadian Wheat Board. This exposure brought home the massive nature of the global wheat industry and the dynamic relationship between players such as the United States, Canada, and Australia in the various global markets.

Then when I left the Foreign Service, I spent a decade seeking biomass for cellulosic ethanol production in Europe, Canada, and the United States. Because wheat straw was the agriculture residue of choice, I had the wonderful opportunity to personally get to know many wheat farmers and, on several occasions, meet with them in their fields. Almost invariably I found closely knit families producing very large quantities of a quality food product by utilizing modern farming practices that, contrary to popular urban opinion, are sustainable with good soil husbandry of prime concern.

When on the topic of wheat, perhaps a few paragraphs on gluten are in order, as somehow this part of the wheat kernel has fallen into disrepute on health grounds. Yes, the 1–2 percent of the population with celiac disease, an autoimmune affliction of the small intestine that is considered to be the result of a particular genetic trait, should definitely avoid all wheat products. There are also people with a wheat allergy, estimated to be less than 10 percent of the population, who can overcome this hardship with standard allergy treatment. Furthermore, less than 1 percent of the population suffers from gluten-sensitive idiopathic neuropathy and also should avoid wheat products. Beyond this part of the population, most of which can find relief from their allergic condition, the remaining nearly nine out of ten people should have no difficulty with gluten in their diets.

Yet somehow a bit of guilt by association has occurred here with the concept that if gluten is unhealthy for some people, it must be an unwise choice for everyone. However, there is no peer-reviewed research published in acclaimed journals that provides any grounds for concern beyond the three conditions mentioned above –which, in the medical context, are considered to be a close parallel to bee stings and peanut allergies. Many food products that contain no wheat components are labeled as gluten free to assist such people. The logic then flows and a perception in the general population arises that if it is to be avoided by some then best to be on the safe side and reduce or eliminate products that contain gluten.

There is a fairly wide spread belief that gluten contributes to autism but research has not supported this concept. However, even if those with this condition were to be included in the population that should avoid gluten, the proportion of people who should shun this common part of our food still remains very much the minority and those who are not diagnosed with one of the inflictions where gluten should be avoided should have no concerns.

It is quite possible that many consumers do not have a full understanding what gluten really is. It is a protein, an essential part of our diet, particularly for those who wish to avoid meat. Thanks to the early domestication of wheat, it has been so since the beginning of organized agriculture some ten thousand years ago. Because it is an exceptionally long and strong molecule, it provides the qualities that make bread rise and form a robust loaf without crumbling. A sandwich would be quite impossible without the latter characteristic. Yes, there are alternative but expensive products such as guar gum that work in a similar fashion, but for 90 percent of the population, gluten should be considered a perfectly healthy source of protein and a standard part of their diet.

Because wheat also contains carbohydrates so there may be further confusion here as the excessive consumption of carbs is known to lead to obesity with the conclusion reached by many is that most or all carbohydrates in the diet should be avoided, and with this point of view this very much includes wheat based products. Moderation in the consumption of carbohydrates at the levels recommended below makes sense. The following is a quote from a research paper by the Food and Nutrition Board of the Institute of Medicine of the National Academies.[118]

> *Adults should get 45 percent to 65 percent of their calories from carbohydrates, 20 percent to 35 percent from fat, and 10 to 35 percent from protein. Acceptable ranges for children are similar to those for adults, except that infants and younger children need a slightly higher proportion of fat (25–40 percent).*

> *The Institute of Medicine recommends 130 grams (520 kilocalories) of carbohydrate per day, which is the average minimal usage of glucose by the brain. The desirable range of carbohydrate intake is 45 to 65 percent of total caloric intake (also referred to as the Acceptable Macronutrient Distribution Range, or AMDR), and the Daily Value (DV) for carbohydrate on food labels is based on a recommended intake of 60 percent of total caloric consumption. These recommendations also generally advise that no more than 25 percent of carbohydrate intake be derived from sugars.*

Sugars are a carbohydrate but are often referred to as providing "empty calories" as there are few additional nutrients that accompany this food. Carbohydrates from wheat are joined by a host of nutrients and fiber and thus make a much better dietary source for this part of our diet.

In summary, wheat is a nutritious, exceptionally versatile, and delicious food that will thrive in climatic conditions too harsh for most other food plants—just what a hungry world needs!

Dairy

My own experience as a dairyman was fleeting but rather intense. When I was about twelve, my father decided that our small dairy herd had to be sold as one of the first steps to consolidate activities and move away from a mixed and uneconomic farming operation. For my spending money, he temporarily provided me with the remaining three cows for the summer to take advantage of the abundant grass, with the provision that I diligently milk them at the same time each morning and evening. Beyond harvesting the milk, I then had to separate the milk from the cream and store the product in a cooler before we delivered it to the nearby creamery for butter production.

This all sounds simple enough, but as with all farming operations the entire operation was surprisingly time consuming. First, the lucky cows were free to roam a fifty-acre pasture, except for the time that they were milked. The majority of the time they showed up near the barn or could be called in from nearby at milking time, but a few times a week they chose to go AWOL. While the pasture was relatively small it was still a half mile long and partially covered with trees, was hilly in places, and had a stream running through it. Thus if the cows were not near the barn, they needed to be located, which occasionally meant a half-hour round trip walk through their known haunts, perhaps to the farthest corner of the pasture and back to the barn.

Free range might be nice for the cows when the grass is lush and the weather is cooperating, but it is a lot more work for the farmer.

For such a small herd, milking was by hand, and to this day I still have somewhat oversized farmer hands—which, I maintain, were the result of this activity. By the time I had collected the cows, confined them in the barn, washed them appropriately, milked all three, let them go, and cleaned any fresh manure, more than an hour had gone by.

The separation of the cream from the skim milk was the next step, which was achieved by using a device appropriately called a cream separator. It was a rather complex device that, through centrifugal action, mechanically divided the cream and skim milk into different streams. At the end of each milking, this device had to be partially dismantled, and the couple dozen components that the milk and cream came in contact with had to be carefully washed along with the pails used in the actual milking. The skim milk was fed to the calves while, as mentioned above, the cream was temporarily placed in cool storage.

For a twelve-year-old the money was quite good, but close to twenty hours of the work was required per week at inconvenient times twice a day, every day to produce enough cream for perhaps forty pounds of butter. As the summer progressed, I developed an appreciation why my father moved from mixed farming to a single food product (beef) and finally achieve economies of scale with a higher income and less hours of work. To further add to the burden of maintaining a small dairy herd beyond my labor, each morning and evening the animals had to be fed over the winter when snow covered their pasture. There was cost such as taxes involved with the land they grazed, upkeep of the barn, fences, and the cream-handling facility plus transport to market. It is based on this experience that I question the viability of small-scale agriculture that is apparently favored by some members of the food-consuming public.

Nevertheless this experience has left me with a lifelong admiration for today's commercial dairy operators, which at least have the advantages of modern equipment, economies of scale, and an opportunity to market the entire harvest from the cow that our hopeless little operation most certainly did not enjoy. Compared to other major farming activities, a modern commercial dairy farm is arguably the most complex of agricultural endeavors. For starters, milk must be harvested at fairly precise times at least twice a day, every day of the year, from substantial numbers of the largest of our domesticated animals. Also as the shelf life of raw milk at room temperature is only a matter of a few hours, rapid cooling is essential after each milking, as is frequent transport in refrigerated tanker trucks to bring the product to a processing facility.

Modern Milking Parlor

The cow, which weighs in at around a half ton, is not necessarily the easiest creature to physically manage. Modern dairy farms have developed systems to handle milk cows with relative ease, including an innovation from the Netherlands whereby the cows voluntarily come in to eat at least twice a day and have robotic udder washing and machines to milk the cows with no human nearby. However, the capital costs of such systems are high, and thus dairy farms need to be reasonably large to effectively capture the economies of scale that such technologies provide.

It was not always this way, and although milk is considered by many nutritionals to be a well-balanced food except to those who are lactose intolerant, over human history it is a relatively recent addition as a daily component of the diet.

Several years ago I watched a documentary on the sinking of the *Titanic*, and a rather inane question was posed to one of the survivors that went something like this: "Other than the unfortunate incident, what do you remember most about the *Titanic*?" The rather surprising response was cold milk. The fellow never had encountered refrigerated milk previously, and because he had the wherewithal to be on the maiden voyage of the most noteworthy passenger liner of the day, he was apparently someone whose family was of some means and exposed to a reasonably high standard of living.

Given that this was about a century ago, this brought me to the realization that widespread beverage milk consumption was not possible in urban centers before refrigerators became a common appliance in most homes.

Also, as mentioned above, the relatively smooth and well-organized milk harvesting and delivery regime of today requires substantial infrastructure, including sophisticated milking parlors, on-farm refrigeration, refrigerated transport, and cold storage (a processing facility where pasteurization, then rapid cooling and packaging, takes place). This is followed by refrigerated delivery to grocery outlets, where the milk is again placed in coolers and presented for sale. It is probable that no other mass-produced food has such a complex and rapid transition from harvest through transport and processing to the point of purchase.

The modern dairy cow is a more docile creature than its earlier cousins of over a century ago. Imagine a less cooperative beast weighing five to seven times that of the farmer patiently putting up with having her privates diddled for ten or so minutes twice a day. Without some means to restrain the animal, the process of milking would normally have been quite an effort and best carried out with only the most docile of the herd. For example, historians give some credence to the fact that Mrs. O'Leary's less than cooperative cow kicked over a lantern and started the Great Chicago Fire of 1871

Another historically wayward cow was that of Laura Secord, who in 1813 during a battle between Canada and the United States, used the uncooperative beast as subterfuge to get away from American soldiers to warn the British and Canadian troops of the enemy's presence.

My own experience, while much less dramatic with restrained animals, still left me with the odd kick in the leg or a dirty foot in a nearly full pail of milk, rendering it totally unfit for consumption.

Given that one could not keep the milk for long and that the ability to transport and distribute were limited or nonexistent, the effort to produce substantial quantities of milk for urban consumption only goes back about a century with the advent of modern refrigeration. The point here is that though milk is a common part of the diet of many people in prosperous societies today, it is not an important element of the diet of our fairly recent ancestors.

The twice or perhaps three times a day harvest is essential for sustainable and large-scale milk production to keep a cow in high-performance lactation for prolonged periods of time, usually about three hundred days from the birth of the calf. While the calf is removed from the mother and placed on a milk replacement ration shortly after birth, consistent milking of the cow is required to maintain the façade that a hungry calf needs feeding at regular intervals. Failure to milk the cow for even for a few hours is not only painful for the animal but triggers a reaction that, if repeated with any frequency, indicates to the mother that the calf is weaning itself and the milk is no longer required. Peak and profitable production levels are quickly lost should this happen.

Then there are other dairy products such as cheese, which have a long history because this product was a means to store many of the nutrients in milk for prolonged periods of time without the need for sophisticated refrigeration. Butter too is another older, mass-consumed product. Given that a modern cow produces milk at about 4 percent butterfat and typically yields about six to seven gallons, or sixty to seventy pounds a day, the actual amount of butter that a cow produces in twenty-four hours is about two and a half pounds. It is only because of the consumer preference for 2 percent and skim milk that retail butter prices can be maintained at such low levels.

Before I leave the section on dairying, the topic of raw milk, which seems to be gaining increased attention and a following in recent years, requires some attention. For those unfamiliar with this issue, all milk purchased in normal commercial outlets is pasteurized, which heats milk to 161 degrees Fahrenheit for 15–20 seconds, effectively destroying all or most pathogens and rendering the product much safer for consumers. Because this impacts the flavor and supposedly the nutritional value of the product, some people prefer the raw product.

I was raised on raw milk, and I prefer the flavor of the pasteurized product, but as I milked the animals by hand with my nose next to the flank, I could always taste what I refer to as "the cow flavor" in the raw version, which may not be noticed by others. Pasteurization and perhaps homogenization (a process that keeps the cream mixed evenly with the milk and prevents it from rising to the top of the container) totally eliminates such a taste for me.

The controversy that arises is that unpasteurized milk is an ideal medium for bacteria that can be deleterious to human health, such as E. coli. Thus health authorities strongly advocate that the public has the protection that pasteurization provides. The very limited number of people following the concept now may not be particularly problematic, as the handful of dairy farmers providing the product tend to practice particularly high standards of sanitation, and the consumers involved understand the risks and treat the product accordingly.

However, for raw milk to become a standard product as just another supermarket offering, there could be some serious health issues. A veterinary friend of mine who specializes in dairy cattle was of the opinion that the standards of both producers and consumers regarding the necessary extra precautions would almost certainly deteriorate and, as stated above, would have the potential to cause a lot of unnecessary illness among the consuming public. It is for this reason that most health authorities will not permit the widespread adoption of raw milk in the commercial food system, which is much to the chagrin of advocates of the practice. My assessment is that health authorities should err on the side of caution, particularly as the nutritional advantages of raw milk over the slightly heated version does not seem to be substantial or backed by any particularly rigorous research.

In summary, milk is a healthy, versatile, high-quality food. However, because of the complexities in its harvest and also the storage challenges that could only be overcome with fairly advanced technologies and infrastructure, it is a newcomer as a major human food.

Beef

During my later years on the farm, we had a feedlot with about three hundred head of feeder cattle at any one time, and thus I have had reasonable exposure to this type of agriculture.

I would like to start with the most positive part of the industry: the cow-calf operation. On land generally with terrain too rugged or dry for other food crops, cows can thrive and produce a calf every spring that gains weight from grass alone and provides high-quality protein for human consumption. In countries such as Australia, New Zealand, and Argentina, where the climate is sometimes blessed with ample rainfall and is invariably milder than all of Canada and much of the United States, the cattle can graze all year, and most reach full maturity on the grass alone. A large quantity of food is produced in a sustainable manner without displacing any other crops and with a reasonably modest carbon footprint. There seem to be few grounds for adversity to such food production.

North America has cold winters in many locations with very limited opportunity for grazing several months of the year, but the abundantly low cost of corn helps balance this out. The several hundred-pound calf, after living its first summer on grass, is then moved to a feedlot, which is a contained area where the only activity is to eat and gain weight until the mature thousand-plus-pound animal is ready for slaughter. To many, this meat is of superior quality to the grass-fed variety as far as flavor and tenderness is concerned. Because of the grain-dominated ration, such beef tends to have more fat distributed in the tissue than is the case for southern hemisphere counterparts that mature on grass.

Those who are against this stage of beef production point out the quality of life of the animal is diminished in such a setting. The methane gas release is very high, which has a vastly more significant impact than carbon dioxide, and it takes many pounds of human quality food to achieve one pound of lean beef.

Cows with Calves on the Open Range

On this last point there is some controversy with those against the feedlot practice, some who claim that up to twelve pounds of corn are required per pound of edible meat. The beef industry advises that it is substantially less. I turn to the Council for Agriculture Science and Technology for the answer as to what is reality.[119] They point out that for a 500-pound calf entering the feedlot and raising it to a mature weight of 1,100 pounds, it requires about three pounds of grain for every pound the animal gains.

There are several reasons for the discrepancies between beef producers and detractors. The most common is the assumption made by those against feedlot beef that the feed given to animals is 100 percent grain for their entire lives. With this assumption, it does take 6.6 pounds of feed for each pound of final weight of the animal, and given that the edible portion of the live weight is approximately 50 percent, their calculations have apparent validity on first glance. However, for the first 500 pounds of the calf's life, there is no grain or other human edible material involved. Also, in the feedlot some of the ration is something other than corn and non-edible for humans as well. Thus taking the entire diet into account over the life of the animal, perhaps each pound of quality, feedlot-finished, lean beef requires about six pounds of edible human food. This is not an ideal situation in the minds of some, but it's certainly not the twelve pounds of human edible material for every pound of beef that opponents industry claim.

Much of hamburger comes from mature beef cows that spent most of their lives eating grass; when around age seven, their productive lives of producing calves comes to a close, and they are dispatched to the abattoir.

The other major source of hamburger is dairy cows, which at a younger age than their beef cousins have passed their peak production years and thus are harvested for their meat. Thus if one is concerned about the high input of otherwise human food for the production of beef, hamburger is by far the better choice. In the supermarket display cases, the high-quality steaks and other cuts that have a marbling effect, with fat dispersed throughout the red meat, generally originates from animals that have matured in feedlots.

Currently the amount of grain going to animal protein production (beef, pork, poultry, eggs, and dairy) seems to be manageable, with corn and other grain producers quite capable of meeting global needs for human consumption. Indeed, many countries which have a twelve month grazing season have a reduced reliance of feedlot rearing of cattle, and it is estimated that 1.4 pounds of human food are required for every pound of edible meat. When one considers the biological value to humans of the superior protein in meat, the nutritional advantage, assuming moderate levels of consumption, compensates for the additional 0.4 pounds of grain required to produce this food.[120] Also on a global basis, humans consume approximately two-thirds of all grains, while livestock consumption is as follows: pigs at 12 percent, dairy cows at 9 percent, beef cattle at 5 percent, meat chickens at 5 percent, and laying hens at 4 percent.[121]

However, consumption of meat and other animal proteins is increasing substantially, and there may be a looming tipping point where the relatively well-to-do, and their preference and willingness to pay for animal protein, may threaten global food security in a serious manner. This issue has been covered extensively in the chapter on the consumption of animal protein.

APPLES

Where I grew up in central Alberta, it is much too cold for the production of apples, although my grandfather had a couple of crab apple trees in a sheltered spot that produced with some regularity. I recall the first time I saw apples grow commercially, was when I traveled with my parents through British Columbia and Washington State to visit my relatives in Seattle. I was quite overwhelmed by entire valleys dedicated to this wonderful fruit. I had not realized at the time that only a good day's drive from my home was perhaps the most prolific apple growing region on earth.

Apples have been a popular temperate climate fruit for perhaps four thousand years (or even longer if one takes the incident in the Garden of Eden into account). In the United States, which accounts for approximately one-third of the world crop, there are an amazing seven thousand named varieties that can be categorized into about four dozen distinct types. The

Pilgrims are credited with introducing apples to the United States in the early seventeenth century, and the French first brought the fruit to Canada at about the same time.

The ubiquitous fresh apple that is always available in countless grocery stores and other food outlets provides a good case study on the very substantial effort to supply a fresh product 365 days a year, although the harvest season is a relatively short period of only a few weeks each fall. Unlike many food products, apples must be harvested by hand, and because bruising creates a serious quality and storage issue, extreme care must be taken by the pickers to collect the fruit in premium condition. The Ontario Ministry of Agriculture and Food goes so far as to recommend that pickers be trained to handle apples like eggs.

Controlled atmosphere apple storage to maintain nearly just-picked freshness for up to a year was pioneered in the late1800s and perfected over the next hundred years. The process is quite sophisticated to apply in practice, but the concept is simple to understand. Low temperatures a few degrees above freezing, appropriate humidity, and the correct balance of low levels of oxygen (plus more concentrated doses of nitrogen and carbon dioxide) in a sealed storage room create an environment that maintain apples and other fruit and produce in a state of near suspended animation, with very little deterioration for many months. Furthermore, this form of food preservation is achieved by simply altering the composition of the earth's natural atmosphere at the right temperature, and thus there are no toxic materials or residues involve.

The largest quantity of apples that most consumers see in any one place is perhaps a few hundred pounds of the fruit in their local grocery store. Few stop to ponder the massive effort that is involved so every consumer in North America has a steady supply of between a half dozen and a dozen different varieties of high-quality apples readily available for purchase all year. According to the US Census bureau, the per capita consumption of fresh apples is close to 19 pounds per year; assuming three apples to a pound, this amounts to about an apple a week. A similar quantity is also consumed in the form of processed apples products, mainly juice.

To demonstrate the scale of this industry, take the greater Denver area, with a population slightly over three million, or 1 percent of the entire US population. To meet the city's requirements, three million apples, or a million pounds of the fruit, must be delivered each week. A million pounds is five hundred tons, or close to twenty semi-trailer truckloads. This means just to keep up to the demand for fresh apples, three truckloads are required per day, which means well over a thousand loads per year for just 1 percent of the population. Furthermore there is an equivalent amount of apple-based product also moving from orchard through processing and then on to consumers. If the citizens of Denver require forty truckloads of apples and apple products per week, the immensity of the food industry becomes apparent when one factors in the many thousand of food items that await not only the inhabitants of the mile-high city but the other 99 percent of the population as well.

Regarding US apple production, this industry is concentrated mostly in Washington State, where only 2,200 commercial growers supply well over 50 percent of the fresh eating apples in the country.[122] Thus each of these farmers, on average, provides enough apples for seventy thousand people. These are mostly owner operated by multigenerational farm families that successfully grow apples on a scale that makes a real difference when it comes to feeding over three hundred million people in the United States. If there were ten times as many growers producing the same amount of apples, what would the advantage be to consumers? The downside would be apple orchard owners with low or even poverty-level incomes and a comparatively inefficient apple collection, storage, and marketing system, with higher prices and probably lower quality fruit as a reflection of growers not having the scale of operation to accommodate the most advanced technology the industry could offer.

To emphasize the reliability of this industry, I ask readers to reflect on when was the last time that they visited a supermarket and there were no apples to be found. It has never happened to me, and I always find abundant quantities of several varieties. I trust readers find it interesting and perhaps useful to have an appreciation of the scale and complexity of the task for all those involved with the movement of an apple from farm to the figurative fork, to meet the daily needs and expectations of hundreds of millions of consumers.

The logistics of making this happen are enormous, and for the supply system to function effectively, each grower must be able to be relied upon to provide the varieties, quantities, and quality required by the fruit wholesalers on dates that are determined months in advance. It is a sad fact of life, but for the trade the cost and administrative overhead is similar no matter what the size of the grower. It is for this reason that smaller orchards with only a few hundred trees are generally passed over by the trade.

To conclude, commercial apple production is most frequently carried out on large family farms offering the consuming public with relatively low-cost and high-quality fruit most of the year, thanks to advanced storage technologies. However, there remains a market niche for smaller growers close to urban centers that have the opportunity to sell directly to consumers at retail prices that are sufficiently attractive to the grower and thus can provide a reasonable return on their investment and labor.

24.

Is the Food We Eat Safe?

In this chapter I will cover both the immediate effects of ingesting tainted food or beverage products as well as including long-term effects on the health of individuals through the excessive consumption of processed foods with high levels of salt, fats, and sugar.

First I will address the more conventional concept of food safety, which is becoming ill or worse from eating food or drinking a beverage that contains pathogens, some chemical, or other serious contaminate. According to the Center for Disease Control in Atlanta, an average of three thousand people in the United States die every year because of the effects of food poisoning.[123] While this is clearly three thousand people too many, it does include those who have consumed improperly stored or prepared foods, where the cause can be traced how food was handled in the home, but not those who become infected on their overseas travels. There is no breakdown on the number that became ill and died because the food was already contaminated at the point of purchase (either grocery store or restaurant), yet the media is quick to pick up on these, and in reality the occurrence does not seem to be particularly significant when compared to other causes of death.

For instance, there were over forty-two thousand motor vehicle deaths per year in the United States and eighteen thousand-plus homicides that occurred in 2007. [124] In reviewing the list of all types of fatalities, food poisoning has one of the lower occurrences of any cause of death. When one considers that the average person ingests a food or a beverage at least a half dozen times a day, the incidence of death on a per-exposure basis is actually extremely rare. I thus question those who maintain that the food that North Americans are offered at the various points of purchase presents a serious threat to overall population health. Regardless, every food-related death is a terrible occurrence, and vigorous effort must be made to minimize the numbers of such fatalities and related illness.

In this regard, there is a legion of inspectors at all three levels of government that, although largely invisible to the public, do a very credible job of keeping our food at the point of purchase at the grocery, restaurant, or fast food outlet relatively safe. Unfortunately, their reach does not go into the home or protect the overseas visitor from the effects of consuming food that is contaminated in either of these settings. And yes, there are lapses and well-publicized unfortunate incidences where multiple deaths and illness occur when contaminated food does make it past the cash register. However, my conclusion is that the food that we purchase is in reality quite safe, and the industry does not deserve the negative reputation that some pundits seem to attribute to it. Everything one does has some risk involved, in comparison to so many other daily activities, consuming food does not seem to be an issue that should concern the public under most circumstances.

It is interesting to note that one of the most common food contaminants is manure, which if left on fresh vegetables as a residue, may contain the bacteria E. coli. Though the vast majority of farmers understand the potential risk and manage around it, this is a downside of organic farming, which has a heavy reliance on manure as a natural fertilizer.

Before moving on to unhealthy food choices, I should point out that food inspection and regulation enforcement do not generally extend to products purchased on the farm or at farmers' markets. Although the folks involved in this trade are generally of the highest integrity and do take the necessary safeguards, the consumer should be aware that the largely invisible protection of food inspection and enforced safety measures are not in place as they are with retail grocery outlets and restaurants.

Unfortunately there is little protection for the consumer to avoid bad choices in their eating habits, and here is where food becomes anything but safe. Because our taste buds respond positively to salt, fat, and sugar, the prepared food industry is providing what consumers ask for in an overabundance. Such products, when they become core dietary items, can be very harmful over time, as has been documented innumerable times in a host of peer-reviewed studies. Excessive consumption of processed foods high in one or more of these three ingredients and lacking in basic nutrition manifests itself in quality of life factors such as obesity and, more significantly, through the increased incidence of fatal diseases such as cancer, heart attacks, or stroke.

My grandmother had an expression about overeating: "A minute in your mouth, a few hours in your stomach, and a lifetime on your hips." This homespun philosophy has a message—there is a very substantial quality of life price to pay for the transient enjoyment of consuming foods that taste good but have limited or no nutritional value and contain ingredients that can contribute to long-term health problems.

As a specific example, my son purchased a half-pound bag of pretzel pieces that were produced by a nationally recognized manufacturer of snack foods. He wolfed down the entire contents in a few minutes and consumed 1,200 calories, or approximately half of his caloric need that day—as well as 146 percent of his daily recommended sodium intake, 75 percent of his fat requirements, and 48 percent of his carbohydrates. On the "positive" side, he consumed 24 percent of his daily fiber and 12 percent of the recommended iron intake, but nothing else. With this combination of ingredients, it would be nearly impossible for him to achieve anything approaching a balanced diet with the remaining calories he had for his day. Ironically the package in rather bold print proclaimed "Zero Trans Fat!" in an obvious pretense by the manufacturers to lull the casual consumer into the belief that the product had nutritional integrity.

I hasten to add that the food-processing industry is equally capable of providing superbly nutritional offerings and does so with a variety of excellent products. In contrast to the snack that my son consumed, a loaf of organic multigrain bread also was purchased as part of our normal grocery shopping that same day. Seven slices of the bread, which weighed the same as the package of pretzel pieces, provided 7 percent of the daily fat requirement, 21 percent of the sodium, 35 percent of the carbohydrates, 70 percent of the fiber, 56 percent of the protein, and between 14 and 48 percent of the daily requirements of eight essential vitamins and minerals. The calorie content for the seven slices was 560, so there was plenty of space in the remaining daily diet to meet all recommended nutritional requirements.

As will be covered in some detail in the chapter on junk food and junk water, it is largely a waste of time to rally against food processors that are doing no more than satisfying the taste buds of consumers who are quite free to choose what they eat. Following consumer wishes is core to the business case of such companies, and they will only change their product mix in favor of more nutritious foods if the market demands such products. The challenge will be to change the consumer to favor the type of product that the bread illustrates and to reject the fat- and salt-laden pretzels. The food processing industry will follow such a trend if and when it will emerge because that is where new market opportunities will exist.

25.

Agriculture in the Third World

For billions of people living in the Third World, the concept of farm to fork is very much different than the system we have in North America. However, as so much of the future of the human race depends on how agricultural developments play out in this mostly tropical setting, I trust that readers will come to appreciate this aspect of global farming.

The topic of food and its production in developing countries could fill volumes given the diversity of climatic conditions, varying states of economic development, differing customs and farming practices, a range of diet preferences, and a host of other issues. Thus it will not be possible in this book to cover so many other issues and delve deeply into actual food production. Rather I will dwell at some length on the often inappropriate attitude and approach the largely temperate-developed countries have to mostly tropical third-world agriculture.

Although imports, and to a lesser extent food aid supply nutrients to much of the Third World, I cannot help but marvel at most counties' ability to feed themselves and how often this capacity is overlooked those from wealthier nations. Take the Haiti earthquake, for instance. Quite naturally I was fascinated by the constant reports regarding the glut of food that was building up at the airport and later in the harbor because the transportation infrastructure was in disarray, and thus delivery to the needy population was an extreme challenge. Every effort was made to get the food to the hungry, and I do not wish to take away from the hard work and dedication by so many to achieve what they did under such difficult conditions. However, what seemed to be missing was an effort to turn to the existing farm community to do what they could to contribute to the cause. Perhaps this happened, but the media's attention was on the Herculean task of food aid distribution effort only, so underreporting may be at play here. As I was not there, I may have erred on some of the following points, but I trust my logic and knowledge of third-world agriculture to portray a possible lost opportunity. Although it was the dry season when the earthquake occurred, there would have been a reasonable crop of fresh vegetables and other staples growing in the fields or stored on farms near the affected population centers. This would have been the basic local rations that would have filled the local markets in happier times.

Common sense dictates that earthquakes destroy buildings but, absent a tsunami, generally leave crops in the fields such as vegetables unaffected. Also farmers most likely live in relatively isolated, single-story dwellings probably built of materials other than cement, leaving them with a survival rate much higher than would be the case for their urban counterparts. Thus, in all likelihood a similar amount of food would have been available as had there been no earthquake and as an added benefit it would have been available from a multitude of sources and locations. Furthermore the local farmers would have been probably more nimble in delivering food over broken roads than the aid agencies, which seemed to rely on trucks.

Unfortunately the economy had collapsed, and the inability by many to pay for the food meant that farmers would have little incentive to market their products. Meanwhile as the internationally donated food distribution efforts became increasingly effective and food was available to the population at no cost, any semblance of a viable local food market deteriorated further. Thus it appeared that local farmers were for the most part ignored as a resource, and their potential to alleviate hunger during the crucial days after the earthquake was never realized.

Evidence that this scenario was a probable reality was an October 4, 2010, Reuters report quoting Oxfam International: *"A massive influx of free foreign food after January's earthquake helped feed displaced people but undercut Haiti's agriculture and hurt farmer's income."*

Also, a few weeks after the earthquake, I read a small article that said the Haitian authorities had requested food aid be discontinued because it was destroying the farm economy; substantial quantities of food was available from local sources. Thus not only was there a viable means to feed the people in the immediate aftermath of the earthquake squandered, but the inflow of food aid had a negative impact on Haiti's ability to feed itself.

Had foreign aid managers first turned to local food sources, there would have been both a windfall for Haitian agriculture and incentive for farmers to overcome any logistics challenges to deliver food to where it was required. All that would have been required by the funding agencies would have been an arrangement to pay cash as food was delivered to prescribed locations. Such a scenario would have also provided a badly needed boost to the moribund economy. Furthermore, the funding for this element of the relief effort would have almost certainly been substantially less than food air freighted in from afar.

As I said earlier, I was not there, and if indeed significant reliance on local farmers by aid agencies was a reality, I compliment those who participated in such a scheme. But if this opportunity was lost, as I suspect, this is an example of the lack of an insufficient understanding of third-world agriculture by those in wealthier nations. I would also like to use this assumed situation to illustrate the dilemma of the entire concept of food aid.

Without question, if a population has serious food security issues through drought, insurrection, or a natural disaster and is lacking the means for commercial purchases, food from international donors is essential, as would have been the case for some nutritional requirements in post-earthquake Haiti. Where the system could be improved would be to include local agriculture as part of the solution. All too often the influx of food aid disrupts or destroys the local market and provides a disincentive for local farmers to maximize production, further exacerbating the situation. While this scenario is well-known to aid agencies, it is usually easier to simply provide the food in a time of emergency rather than spending the time and effort to understand the local agricultural scene to determine how best the farming community could be mobilized. Unfortunately, the lack of expertise on the dynamics of third-world agriculture, which varies substantially from country to country, is lacking in most donor agencies.

I have firsthand experience in such a situation. When I was posted to Thailand, the Canadian International Development Agency had funded a dairy project that was not meeting milk output expectations. I had more agricultural experience than others in the embassy, and in spite of having no previous exposure to either tropical or third-world agriculture, I was dispatched to the project to determine why milk production was unacceptably low. In this instance there was unhelpful speculation that the farm workers had a private milk marketing business on the side and were diverting part of the production for such an enterprise.

Upon arrival I immediately understood the reason for the disappointing milk output. The breed of cows were Holsteins and were imported from Canada, where they were bred to thrive in rather harsh winter and temperate conditions in summer. In Thailand the coldest month is January, which still has a daytime high of eighty degrees; the rest of the year is even hotter with almost constant high humidity and very wet conditions in the rainy season. Thus the cows were gaunt and clearly stressed from such an unrelenting tropical climate. Furthermore, the rations consisted largely of cassava (a starchy root), broken rice, and some hay that was a tropical, rather course, grass quite different than one might find in North America. This was an inappropriate ration for such a breed of cattle, and it further led to their lack of productivity.

The fact that a lot of money was spent on the assumption that a complex temperate agricultural concept could be transplanted from Canada to rural Thailand and be successful illustrates that aid officials, while well meaning, often are not equipped to grasp even the fundamentals of farming in the Third World, particularly in a tropical setting. As readers will recall from the section on dairying, this is the most sophisticated of farming endeavors given the twice daily harvest of a product that is highly perishable without refrigeration, and it is therefore critically dependant on a reliable power supply, which was also an issue at that time in rural Thailand .

Though milk is a nutritious food, it is not a natural food for humans beyond two years of age and therefore could hardly be considered as a priority product for the Thais, who in many instances are lactose intolerant. In addition, Thailand has diverse and well-developed tropical agriculture, and it is more than capable of consistently meeting its own food needs with an array of nutritious products. Indeed the country is a substantial exporter of a host of quality food products that are readily available in most North American supermarkets.

I hasten to add that I do not consider myself particularly knowledgeable regarding third-world and tropical agriculture, but I am far enough up the learning curve to know what I do not know on the topic. For example, I have no idea how to manage a successful dairy operation in such a setting, but I do know something about temperate agriculture, which is often of little relevance in such situations. In this regard, my experience has been that aid agencies around the world, which are almost invariably from developed countries with temperate climates, do not have the depth of tropical agricultural expertise to adequately provide the research, guidance, and sound project management from which the Third World's farmers can effectively profit. Certainly there are exceptions such as the American Norman Borlaug, who is credited with great advances in wheat genetics for developing countries and is sometimes credited with being the father of the green revolution.

My point is that there is a critical shortage of such expertise, and it may be the reason why agricultural aid has fallen dramatically in the past generation. To illustrate this situation ,I draw from a couple paragraphs from a paper entitled "Trends in Agricultural Aid," prepared by Trinity College, Dublin.[125]

> *Official Development Assistance (ODA) to agriculture decreased in real terms by nearly half between 1980 and 2005, despite an increase of 250% in total ODA commitments over the same period. The share of ODA to agriculture fell from about 17% in the early 1980s to a low of 3% in 2005. In Sub-Saharan Africa the reduction in agricultural aid was less dramatic, but still sizeable, with a decline of about 35% over the period.*
>
> *Bilateral aid to agriculture has also fallen from 12% of total bilateral aid in 1980–81 to 6% of the total in 2000-01. For individual donors, the fall is even more striking. For Canada, the fall was from 22 to 4%; for New Zealand, from 25 to 3%; for the Netherlands, from 21 to 3% and for the United States from 18 to 4%. The multilateral institutions have also reduced aid flows to agriculture, from 35% of their total flows in 1980–81 to 7% in 2000–01.*

I suspect that the dairy project that I encountered in Thailand was more the rule than the exception, and thus managers in the various aid agencies quite naturally focused on projects with a much greater chance of success that were outside of the agricultural sector.

Before I leave this chapter, I have some concern that in North America, where there are many personalized approaches to food that are perfectly fine for the individual enjoying a prosperous lifestyle, well-meaning but ill-informed individuals might attempt to transport these to the Third World. In my opinion most such food sourcing and production preferences have little relevance in such a setting. Indeed, if the trained agricultural specialists in developed countries struggle to understand the situation in the Third World, the disruption in food production that amateurs would cause could be serious.

For example, the anti-GMO movement has been able to convince at least two African governments to reject major deliveries of food aid because of the claim that the products were contaminated by genetically modified grains and therefore were potentially toxic. Although this was a victory to those living in the comfort of plenty, they seemed to overlook the fact that this food was needed to feed hungry people. The downside of accepting this most insignificant of risks (the food was the same as North Americans regularly eat) was the very real potential of hunger, malnutrition, or possibly even starvation.

I have written a fair bit on what many of us do not know or appreciate about third-world food production, and now I am going to stick my neck out an make a few observations on how food production is going to evolve in several countries falling into this general category.

I'll start with Brazil, which by most standards has largely graduated from Third World status, particularly regarding its agricultural sector. The farming industry here has made some very impressive gains in recent years, to the point that in 2009 it was the global export leader in beef, chicken, orange juice, and coffee, and in second place with corn and soybeans. While many consider any agricultural expansion of food production anywhere in the country to be at the expense of the rain forest, such growth is almost all in the Cerrado[126] or the prairie-savanna, which is the previously relatively unproductive but immense central swath of the country approximately the size of the US and Canadian Great Plains combined.

Until the past decade the land was considered unfit for intensive agriculture because of the soil acidity and the tropical climate, which precluded the production of most of the temperate food crops for which there is a global market. By adding substantial quantities of lime to reduce the acidity, tropicalizing such temperate crops as soybeans, and enabling two crops to be produced each year in such a setting with a twelve-month growing season, yields were doubled in comparison to almost everywhere else that soybeans are produced where the climate is suitable for only a few months of the year.

Though the intensive commercial farming of this great region has no impact on the rain forest, there are environmental consequences. However, the amount of food produced to feed an ever growing global population facing climate change is perhaps worth the trade-off. Furthermore Brazil has emerged, almost exclusively through homegrown ingenuity, as the world leader in advanced tropical agriculture technologies. A real opportunity now exists for this knowhow to be exported to the Third World with similar climates such as Mozambique which has initiated such an arrangement with Brazil[127]. As the Brazilian approach is much more relevant in such a setting than that which has been tried from temperate agriculture, it has good potential to assist such countries to reach their food-producing capacity.

Sub-Sahara Africa, with the exception of much of South Africa, is in many respects similar to Brazil in overall land mass, climate, and large areas of open savannah. Unfortunately unlike Brazil which has enjoyed good governance in recent decades, most countries in this region lack the infrastructure and economic wherewithal to emulate this model. Much of the agriculture is small-scale subsistence farming, which is both labor intensive and inefficient with overall food production, falling far short of the potential. Though environmental sensitivities, particularly given the wildlife, are an important consideration, there is a major untapped source of badly needed additional food that the world will require to feed the increasing local and global population.

Assuming that the agriculture technology that is applicable to this region is available from Brazil or possibly elsewhere, other factors must fall into place before real progress toward commercial agriculture can take place. The first is stable and presumably democratically elected governments, the likes of which provided Brazil with the ability to succeed on many fronts. This will lead to enhanced economic development that will provide more meaningful employment off the land for many of the subsistence farmers. There may be some nostalgic considerations for the herdsmen with a few animals and those with small plots of grain or vegetables, but mostly they lead a marginal existence with extremely limited standards of living and little ability to improve the lives of their children.

If there is actual property ownership in some rural areas in Sub-Sahara Africa which is unfortunately often not the case, there is then an asset to sell, providing the opportunity to move to a reasonable job elsewhere. Not only will this provide an easier transition to a different lifestyle, but the more prosperous and perhaps better farmers will be in a position to increase their holdings and move toward a commercial-scale operation with better returns for their efforts and almost certainly increased food production from the available land. This is much the same scenario as the transition that North American agriculture went through during much of the twentieth century, and it continues to this day. This will not come early or quickly in most countries, but given the potential this region has to offer, the incremental food production that would occur would be a most welcome addition to ever growing global requirements.

There is another phenomenon that may have a significant impact on food production, particularly in Sub-Sahara Africa. In recent years, particularly the Chinese are reportedly purchasing large tracts of land for food production.[128] There is some evidence that the practice is not as widespread as some members of the media report, but the concept is troubling. The rapid introduction of large commercial farms amid subsistence agriculture has many pitfalls.

First, the accumulation of large tracts is often only possible if existing farmers are displaced. Given weak or non existent land tenure legislation (or the necessary enforcement should such laws exist), there is a very real potential that the existing small farmers are displaced without meaningful compensation for land that might have been held by the family for generations. Furthermore the opportunity for a displaced farmer without evident skills, who is suddenly on the labor market against his or her wishes, may be practically nonexistent, particularly as unemployment in rural areas tends to be high. Also, with the advent of mechanization, the newly established farms are unlikely to require a labor force approaching the number of displaced farmers.

Countries such as China have their own food security issues and will be inclined to look upon such farming operation as a guaranteed offshore source, with no interest in the local market. This in turn could cause at least local food shortages. Alternatively, if the market of interest was local, such farming operations—with the advantages of the efficiencies of mechanization, scale, and advanced technology—produces food at considerably lower costs than the remaining subsistence farmers. If this cost advantage manifests itself in inexpensive food in the local marketplace, although it's good for consumers, it could be devastating for the remaining traditional farmers.

In North America where farm consolidation was a gradual, multigenerational phenomenon with few new players entering the rural scene, the transition to large commercial farms occurred with a relatively low level of economic or social disruption. Foreign land buyers suddenly changing the dynamic of whole communities, in what some writers refer to as a land grab, is a very different and troubling phenomenon. Both the FAO and the World Bank are monitoring the situation, and if it persists or becomes an accelerated trend, I expect that these and other like-minded organizations will make an effort to oppose or at least encourage some level of control to mitigate the disruptions discussed previously.

Although China is a leader in offshore land acquisition, this country is also the largest food producer on the planet. I know of no other country which has such a contrast in all aspects of food and agriculture. Starting with the up-market food outlets in Shanghai, Beijing, and many other major cities, these rival anything in variety and quality that one would encounter in North America and Europe. However, in rural areas, particularly the further one goes into the interior, the markets are closer to what one expects in a third-world setting and offer a much more limited choice dominated by locally grown products.

The diversity in agriculture is even greater. With Siberian-like conditions to the north, and the tropics of Yunnan Province and Hainan Island, one can find practically every type of farming that the entire world has to offer. What is equally striking is the socioeconomic conditions found in rural China compared to their urban counterparts, particularly on the coast. Though there are some modern, large-scale commercial farms with the technical capability to produce food with the efficiency of the Brazilians, most farming is on very small plots with intensive labor inputs.

The following photo depicts a typical farming operation in China. The plot of land that is in the photo is approaching an acre and represents over a half of the average holding in rural areas of the country. There are five farm workers involved with the harvest on a field the size of which the North American farmer on a large combine would have completed in a matter of minutes. The quality and nutritional attributes of the grain from both scales of operation would be nearly identical, but the mechanized farmer sits in a cab with all the comforts of a late model car while the Chinese peasants toil in much less comfortable conditions for a tiny fraction of the output per person. Thus compensation per hour of toil would be minimal.

Labor Intensive Harvesting in China

China has the technology and often the capital to modernize their agriculture, but to do so would not be possible until alternative employment is available for the masses of peasant farmers. Though there is an inefficient use of labor, these farmers do a very impressive job of feeding close to 20 percent of the global population on only 10 percent of the arable land.

This scenario has a serious downside, with a profound impact on the possible future of China. According to 2011 statistics on China as prepared by the CIA, 39.5 percent[129] of the labor force of 820 million remains engaged in agriculture. To put these large numbers into perspective, the Chinese have the equivalent of the entire population of the United States dedicated to feeding the country. The income distribution for agriculture poses the real problem and is frequently overlooked by pundits who are enamored by the impressive growth of the manufacturing and service sector. Again according to CIA statistics, agriculture's share of GDP was only 9.6 percent.[130] In round numbers, the 40 percent of the labor force engaged in agriculture are rewarded with 10 percent of the total GNP. On an individual level, this means that an individual farmer in China earns about the equivalent of two thousand US dollars per year, whereas those engaged elsewhere in the economy have an average annual income of over ten thousand.

This dynamic presents a huge challenge for the future of China because it is extremely dependant on the continued willingness of the peasant class to toil at low levels of compensation to feed comparatively wealthy citizens. From earlier firsthand observations when I lived in China, and based on my continued interest in this situation, it is my opinion it will be food, and those responsible for its production, that will have a very major impact on the future of China as a great power. Should the rural food-producing underclass become sufficiently dissatisfied with their lot, become a coordinated force, and simply choose to withhold some of their product from the marketplace for even a short period of time, the disruption that it would cause in urban centers would be substantial, perhaps to the point that it would threaten the very pillars of the communist establishment.

However, there are some positive steps undertaken by the central government in recent years to mitigate this imbalance. The major initiative was land reform. Initially all land belonged to the state, and through family tradition farmers had access to the plot that their parents and previous generations had tended. Thus as long as one generation took over from the next, tenure was reasonably secured. This provided a huge disincentive to leave the farm even if incomes were low, because forfeiture of the land to the state—or more likely corrupt officials—would probably have been the result. If there was more than one son in the days before family planning, plots of land became even smaller and less efficient. With title, a farmer can rent the land or sell it to provide themselves with funds to set up a business or transition into a job in the non-agricultural economy. The added advantage was that land consolidation could occur with the remaining farmers acquiring larger holdings that were suitable for less labor-intensive and more productive, advanced agricultural practices.

China can only hope to become a fully modern nation, without the potential of unrest in rural areas and a disruption in food supplies, if an orderly and reasonably rapid transition is made to an agriculture model closer to that of the United States, where about 1.5 percent of the labor force,[131] not 40 percent, feed the nation. It is interesting that food, or more precisely the producers of food in China, will have so much influence on how that country evolves into a modern global power.

Before I leave this chapter, I wish to share a personal observation on tropical rain forest degradation and food production. A few years ago my wife and I spent a few days on an eco tour downstream from Iquitos in Peru, on the

Amazon River. In that very verdant place, guides took us to several locations slightly inland from the river to demonstrate the effects of slash and burn and what the Peruvian government was doing to counter this misadventure by local farmers.

First we visited an old-growth forest that had not been disturbed and remained as pristine as would have been the case in pre-Columbian times. Then we went on to an area that had experienced slash and burn but had been abandoned by the farm community which had moved en masse to a new location. Although the area of perhaps a hundred acres had only been deserted for three or so years, I was amazed by the size, quantity, and variety of the vegetation Bird life seemed abundant, and although there was no opportunity to explore for animals and reptiles, they probably were also present because the lush habitat seemed suitable for all types of fauna. What occurred to me was that while slash-and-burn agriculture was clearly devastating to mature jungle, the comings and goings of primitive farmers on areas that had been logged may not be as environmentally disruptive as some writers and others make it out to be. Regeneration of the jungle including sizable trees in just a few years in a high-rainfall, twelve-month growing season is amazingly rapid, with renewed carbon sequestration apparently occurring at an impressive pace.

Later we visited a small village located directly on the Amazon, where agricultural produce was being loaded for boat transport to Iquitos, the nearest substantial population center. We walked a few hundred yards past the village and encountered a fairly recently installed cement sidewalk about five feet wide running through the jungle. There was a busy run of three-wheeled bike transport with loads of fresh fruit, vegetables, small animals, poultry, and charcoal destined for the nearby boat landing. It was apparent that prior to the sidewalk's construction, all produce had to be carried in small quantities on the backs of humans or by pack animals that were expensive to maintain. This most primitive of transport was now replaced by simple pedal-powered tricycles with a laden cart in tow. Here a basic walkway had a very major impact on the quantity of produce that could be moved from the fields to boat transportation.

Besides the economic benefits to the local farmers, this vastly improved transport link tied them to the adjacent land. It now was much more profitable to practice sustainable growing methods on the same plot of land with ready access to the established infrastructure than to move elsewhere in the jungle, where the yield advantage of a new land after a slash-and-burn exercise would not offset the advantage of the ease of transport of their produce to market. Such a basic infrastructure solution, and such an impressive win for both the farmers and the rainforest!

To summarize, agriculture in developing countries is in a state of transition, particularly with those jurisdictions that have reasonable governance. These countries are enjoying sustained and strong economic growth, and thus meaningful job possibilities for subsistence farmers, which leads to land consolidation and increase food production. This is much the same trend as parts of North America have experienced in recent decades. In contrast to this, for those countries that are less fortunate on the governance and the economic development fronts, subsistence agriculture with its underemployment will prevail. However, over the next decade the population of farmers in the Third World can be expected to decrease substantially as the underpaid farm workers move to more productive activities elsewhere, and the food output from this group of countries can be expected to increase substantially as land consolidations leads to a scale of operation that permits advanced agricultural technologies.

Before leaving this chapter, I ask readers to compare the planting operations in the two photos on the following page. In one there are six third-world farmers planting by hand with an output that would be no more than a few thousand square feet per day. This is backbreaking work resulting in a very modest output beyond what is required to feed the family. In the second picture two tractor operators will each plant a few hundred acres per day with the assistance of a truck driver, who will keep the seeding equipment supplied with fertilizer and seed. Considering that an acre is forty-three thousand square feet and would require the effort of the six workers perhaps a couple days to plant, the mechanized approach to farming results in perhaps a thousand fold increase of planting with the same number of workers. Thus until such time as farm consolidation takes place in the Third World, the returns on labor will be insignificant and which will mean persistent rural poverty.

Labor Intensive Rice Planting in Asia

Mechanized Planting of Grain in North America

26.

Food Security, Insecurity, and Famines

The journey of food from farm to fork is by no means a perfect one for everyone, no matter where he or she lives. Food insecurity is a reality to varying degrees in virtually every country, including the United States and Canada.

According to the United Nation's Food and Agriculture Organization (FAO), there are some two hundred definitions for food security and insecurity, all of which are close variations of the same theme. For purposes of this chapter, I turn to the following definition, which is one frequently used by the FAO.

> *"Food security exists when all people, at all times, have physical, social and economic access to sufficient, safe and nutritious food which meets their dietary needs and food preferences for an active and healthy life. Household food security is the application of this concept to the family level, with individuals within households as the focus of concern.*
>
> *Food insecurity exists when people do not have adequate physical, social or economic access to food as defined above."*

Many readers may find it surprising that the USDA, which tracks such issues, has determined that in 2009 some 85 percent of all households were food secure, and slightly less than 15 percent were food insecure.[132] This means that close to forty-five million Americans, in such a land of plenty, do not have enough food at least on some occasions. There is an element of malnutrition and perhaps unfulfilled hunger for at least part of the time for some folks (9.0 percent of the total population) and, for others, (5.7 percent) most of the time it is a near constant struggle to secure food needed to sustain themselves, which means there are occasions when there is real hunger. This is far from a problem that is going away, as the 2009 food-insecure population was at an all time high.[133]

Readers who favor the hundred-mile diet, organic foods, have concerns about GMO foods, the lack of heritage varieties, have a preference for organic, or are against farm-raised fish should reflect on the good fortune to have the luxury to entertain such concerns or make such food choices and consider the plight of all those with food insecurity. I would hope that this very serious situation of close to one in seven people gives meaning to the very real priority of farmers and all aspects of the food chain to deliver low-cost and nutritious food to society in the United States and Canada, where food insecurity is also an issue of similar proportions.

On a personal note, my wife and I are strong supporters of the local Food Bank movement, and we routinely purchase a large bag of rice and leave it in their donation box at our local grocery store. Also, when we have a house party, we ask our guess to bring a bag of rice or other non-perishable food product in lieu of a hostess gift. Our friends tend to be generous lot and still bring the customary bottle of wine or two along with their food donation. In addition some have adopted this practice and also collect for the Food Bank, which does an excellent job of reducing food insecurity at the local level. I invite readers to consider supporting their nearest such institution.

Globally, food insecurity affects nearly twenty times this number, and probably in situations much more dire than are faced by food insecure North Americans. In a 2008 comprehensive study of seventy developing countries by the USDA, it was concluded that 819 million people in the Third World faced food security issues, with most in Asia and in Sub-Sahara Africa.[134] This global problem is a permanent feature in most populations but with varying degrees of severity. Furthermore, in the next ten years the global food-insecure population is expected to increase somewhat to 834 million,[135] or approaching three times the population of the United States. During this decade it is expected that the Asian situation will improve significantly, but the number in Sub-Sahara Africa will increase to nearly half a billion, which is a fairly clear indication that modernization of agriculture in this part of the world is not expected to occur anytime soon. Close to double the US population in one region facing food insecurity is, to me, perhaps the most serious global issue of the early twenty-first century.

Beyond food insecurity there is famine where the entire population of a region is subject to extreme food shortages over a prolonged period of time with the possibility of wide spread mortality due to starvation. As horrible as famines are they are actually impact only a very small proportion of the global population. It is important to note that they are not the result of a global food shortage but rather a failed state that can not, or for political reasons, will not care for its citizens such as is the case with Somalia in 2011. Besides challenged governance which can hinder outside food aid, a local crop failure combined with inadequate infrastructure such as roads and storage facilities all exacerbate an already desperate situation.

Amartya Sen, the Indian economist and winner of the 1998 Nobel Prize in Economic Sciences made the observation *"that there has never been a famine in a working democracy"*. This profound observation emphasizes the negative role such authoritarian states as North Korea play in the welfare of their population.

Returning to North America and our food security issues, the 85 percent without such concerns are indeed fortunate, but they should not be totally complacent because the food production capacity of this continent and elsewhere is perhaps more fragile than many would imagine. Although it would be a highly unlikely event, if a cold front covering western Canada and extending down into the Great Plains as far as central Kansas and east to Iowa brought the temperature down to twenty-seven degrees Fahrenheit for one hour in the wee hours of an early June morning, North America (and indeed the world) would have an instantaneous and huge food insecurity issue because the much of the continent's wheat, barley, oats, corn, soy beans, and other important food crops would suffer such frost damage that they would not recover. Most fields in one of the world's greatest breadbaskets would be lying fallow or, at best, be marginally productive until the harvest of the following year, some fourteen months later as it would be too late in the season to mobilize a massive, unexpected seeding exercise of fast-maturing crops.

These crops and the livestock that they support provide a very significant proportion of both the nation's daily food intake and the products for the export market. The absence of such a large proportion of the available food supply would cause widespread shortages, extreme food prices, and a probable return of even more violent food riots that much of the world experienced in 2008, when it was Australia (and a much smaller crop shortfall) that caused the disruption to global food availability. This particular scenario of mass food insecurity is perhaps farfetched, but the ripple effect of any significant food disruption from any major producing region is of a global consequence, as was the case in 2010 when Russia banned the export of wheat because of a poor harvest, and the price of this staple increased substantially with unfortunate consequences for the less fortunate in this world. Indeed, food prices were a significant contributor to the turmoil in Tunisia in early 2011, which in turn had profound consequences elsewhere.

I am fortunate because I have never faced food insecurity in my life, where there was no option but to remain hungry for a significant period of time without ready access to the next meal. However, I once encountered the very real possibility of such a situation on a personal level for myself and my family, and it was a most uncomfortable experience. In the days leading up to the Tiananmen Square incident, there was every indication that the entire country was becoming increasingly dysfunctional, and thus the massive effort to keep food flowing from the countryside for the approximately fifteen million inhabitants of Beijing, where I lived at the time, was becoming an increasingly challenged undertaking. In addition to the ever escalating civil unrest, there was some expectation that the protesters would prevail, and as there were no indications there were participants in the movement capable of forming a government even remotely equipped to manage a country of 1.3 billion people, a severe disruption of society and economic activity was a looming possibility. At that time it was the general consensus in the diplomatic community that the city of Beijing normally had an available food supply positioned within the metropolitan area capable of sustaining the population for about a week before shortages became evident. Thus the specter of hunger as a consequence of the turmoil loomed over the city.

At that time I was living about three kilometers from Tiananmen Square, and thus I was very near the epicenter of the strife. My household consisted of my wife, one adult son, three smaller children including a week-old baby, and my mother-in-law—all in all, a substantial daily food requirement. In anticipation of the worst, I set out to stock up on some basics from a Chinese outlet reserved for diplomats. When I arrived a truck from the then Soviet Union embassy had just completed loading what would have been a couple tons or more of rice, water, and canned goods, so it was apparent that others were equally concerned. The observation of an acquisition of this magnitude led me to fully load the back and front seat and trunk of my car with several hundred pounds of a variety of mostly staple, non-perishable products; I was a hoarder for the first time in my life. Undoubtedly many thousands of others, Chinese and Westerners alike, were doing the same, so the food security situation became even graver to those who did not stock up as the limited food supply in the shops throughout Beijing was fast being depleted.

I took my precious cargo home and then began to worry how I would protect it if a hungry mob raided our up-market apartment block, which signaled prosperity and probable stocks of food. I was therefore personally relieved when the government crackdown of the protest movement occurred and the danger of food shortages quickly subsided.

This is by no means an isolated incident, but hoarding by those able to do so is a natural consequence whenever access to abundant food stocks is threatened. This human tendency can quickly make even the *concern* about shortages become a catastrophic food insecurity reality for the more vulnerable in society.

I have one more food security situation before I return to the concept of what governments are doing about the issue. This involves my mother-in-law, who came of age in and around Hamburg during the Second World War. To this day, she methodically runs her finger around the inner shell of a recently cracked egg to glean the small amount of the white, maintaining that this provides the equivalent of an extra egg out of every dozen. This is one measure that desperate, food-insecure Germans faced during the later stages of the War before the Allies started to help them reassemble the badly broken country.

In response to this situation, the European Union (EU)—or rather its earlier incarnation, the European Economic Community—established the much maligned Common Agricultural Policy (CAP).[136] I use the word maligned because the critics are against tax payers' dollars being paid to farmers creating what appears to be surpluses, which are then sold at world prices that are generally much lower than the costs of production. The intent of the CAP is to create and maintain a robust and stable farm sector so the specter of widespread European food insecurity never becomes an issue. This of course is lost on the critics who assume that abundant and ever present food supplies is a fundamental and perpetual bounty of the land that occurs automatically. The policy makers and politicians in the EU do not elaborate on this fundamental objective of the CAP because they probably do not wish to unduly concern the public that food security could become an issue.

Similarly the equally maligned US farm safety net program[137] faces criticism and precious little praise for keeping the supermarkets well stocked so that at least the majority of the population does not even have to contemplate food security issues. For most Americans, this perpetual abundance—and at a cost that is available at only a small fraction of disposable income—is a natural as the air they breathe, and this most fundamental of needs never becomes a significant issue that confronts society. I pity the poor president and other elected officials if ever widespread food insecurity was to occur.

So-called surpluses are constantly required as a buffer against the vagaries of harvests to provide confidence in the marketplace that shortfalls are not likely to occur. This confidence can easily be shattered, as was evident in 2008 with the drought in Australia. The shortage was in reality only a temporary lower than average global stockpile of one grain, wheat, and yet the ripple effect across all grains was profound. In terms of total food available in comparison to the actual global needs, there was always quite enough to feed everyone. However, the specter of less than abundance, and hence the possibility of shortages, sent not only the markets surging but countries such as India placing an embargo on rice exports, which is a form of hoarding and thus further exacerbates the situation.

Even at the typical household level, there are food surpluses in the sense that there is normally a few days' supply of at least staples and probably an abundance of a host of products. Maintaining a larder may not only be a convenience when it is time to feed the family, but it could easily also be a subconscious act to foster food security, which has been practiced from the days of the hunter-gatherer. Confidence that the next few day's meals are at hand provides a necessary level of reassurance. The same dynamic exists on the household level as on the global food supply, where the comfort level is a surplus of 20–25 percent of annual global production or the equivalent of all harvests for about three months. The FAO and others monitor the global food supply constantly because this is an extremely important indicator regarding the food security plight for hundreds of millions.

Most citizens of North America—and, I suspect, many elected officials—do not give food security or insecurity a second thought and assume that supermarkets will be places of abundance. However, the consequence of actual and wide spread food shortages to the point of actual famine on a global scale would arguably be the worst of any calamity that could befall mankind. Masses of hungry people are perhaps the greatest threat to the fabric of society, as was evident when the wheat shortage occurred and some governments fell simply because prices became uncomfortably high—yet in spite of the civil strife in these instances, real food shortages never did actually materialize.

And yet why is there a persistent incidence of food insecurity, even in wealthy nations with a consistent abundant food supply? At the individual level it is almost invariably a socioeconomic issue that includes such factors as the lack of funds to purchase food, as well as the infirm who do not have the mobility to acquire their groceries or perhaps those who reside in a "food desert," usually defined a low income urban area that is not served by retail outlets offering a normal range of fresh fruits, vegetables, and other nutritious foods.

However, instead of taking the secure food supply situation that we have for granted, or even dismissing it as broken as some are inclined to do, I trust readers will have more respect and gratitude for the system that is in place today. For those who find all sorts of reasons to condemn many farming practices and yet provide little in the way of alternatives that would still deliver food security to so many billions of people globally, perhaps they can at least give this very complex sector some credit for their perpetual ability (so far) to do an adequate job of feeding the world.

I say "so far" in the above paragraph for a good reason. Experts on the global food supply, such as those in the FAO, are becoming increasingly concerned that the global stocks, measured in days at current consumption rates before they are depleted, are becoming increasingly worrisome. To me, this is even more reason to encourage the adoption of advanced farming practices throughout the world if mankind is to avoid increased incidences of food insecurity and actual famine.

27.

Abuse of Food in the Kitchen

Many individuals express concern about the food system and have a perception that it is somehow failing on several fronts, but they may not realize that they can take matters into their own hands on the later part of the journey from farm to fork. Food is wasted or otherwise abused, albeit probably unwittingly, by many consumers. They either pass up the opportunity to purchase vastly superior foods that are readily available in favor of the nutritionally challenged processed varieties, or they start out in the kitchen with fresh and natural fruits and vegetables but destroy much of the nutritional value and flavor by altering its state through peeling or cooking, essentially developing a processed substitute for the original item in much the same manner that the food industry is castigated for doing.

Arguably the most abused food is wheat, which is an important part of most North Americans' diet and accounts for 20 percent of global calorie intake plus significant nutrition to boot. Here is what happens. The farmer delivers a bushel (sixty pounds of wheat) to the market, where a relatively small proportion is simply ground into whole wheat flour. The bushel is sufficient for the production of ninety loaves of whole wheat bread, which in most supermarkets accounts for only a relatively small proportion of the shelf space allocated to bread. Unfortunately, because for some reason most people prefer white bread, abuse sets in not only with the purchasing decision but with the types of flour normally purchased for food preparation at home.

The abuse is twofold: loss of both quality and quantity of this most basic of food. To get white flour, one-third of the bushel is essentially discarded and goes to livestock feed, leaving sufficient quantities for only sixty loaves of white bread. The part that is discarded contains the bulk of nutrients and fiber while leaving mostly carbohydrates.

Without getting too technical, the seed has essentially three parts. The outer shell, which is usually a reddish brown, is mostly fiber that we need in our diet, but because this is supposedly an offending color, out it goes in the milling process (it is feed for cattle, essentially converted into a steak in a feedlot). Then there is the part of the seed that brings life when combined with water and warmth the following spring. As this too is brown, it also ends up as cattle feed. Given that this part holds the secret of life, it is not surprising that complex nutrients are concentrated here. The remaining two-thirds of the seed is mostly a carbohydrate, which provides the food for the newly sprouted seed to facilitate the growth of roots and the first leaf. It is colorless, and it is deemed suitable for white flour and the bulk of bread and bakery products that are cherished by consumers.

The rather stark loss of nutrients resulting from the milling of white flour is clearly illustrated in the following graph, for which I have to thank George Yu, author of the excellent book *The Simple and Natural Way to Vibrant Health*[138]. He complied this from the referenced USDA data source, and because his book focuses on health, he has provided a much more detailed account of the nutritional consequences of this issue than I provide here. However, the substantial nutritional advantage of whole wheat products over the white alternative is readily apparent with one glance at this graph.

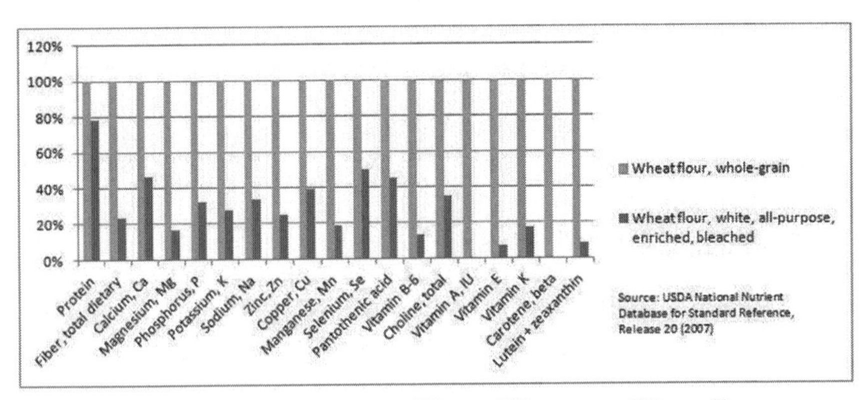

Nutritional Comparison of Whole Wheat and White Flour
Permission granted by George Yu

Things do not end here. Freshly milled flour is not pure white and only becomes so after aging; thus to facilitate the whitening process, millers add bleaching agents. Considering the potential effects on human health of bleaching, this practice is quite rightly subject to some controversy. Indeed the Europeans have deemed that the process is potentially detrimental and is of limited advantage to consumers and thus the EU has banned the use and sale of bleached flour. I would view a similar prohibition in North America as a positive development for consumers without any significant negative impact for millers and bakers.

Steps have been taken to alleviate the recognized nutritional deficiency of white flour in Canada and the United States, and thus this product is "enriched" with at least B vitamins and iron. There seems to be a gap in logic associated with this issue given the need to go to this end just for the perceived cosmetic benefit of white flour and the products thereof. As our household uses mostly whole wheat flour and products, my wife and I have long had a taste preference for the brown versions.

A pragmatic development is the contribution by plant breeders and a willing farming industry to grow a white variety of wheat. This type, when milled, produces whole grain white flour with all the nutritional benefits of the conventional brown version. Interestingly, my adult children, who I must admit prefer white bread but seldom are offered this choice, assume the products that come into the home made from the whole white wheat are a lapse in our normal purchasing habits. One cannot help but wonder, does white in this instance have anything to do with flavor?

Some might claim that this preference for white over brown has nothing to do with the kitchen but is the result of what the food industry foists on hapless consumers. As stated elsewhere in this book, consumers set demands and industry in the free market economy in which we live will follow to maximize sales. In the case of whole wheat bread and flour, the same companies provide both, so here there is no brand identity, and the profit would be similar for both. The superior nutritional attributes of whole wheat products should be known to those focusing on healthy eating, so why is society so willing to accept such an inferior product? To emphasize my point, I recommend that when next readers are in a supermarket, they do a quick comparison on the shelf space allocated to whole wheat bread (not simply brown bread, which, though better than white, is not made from the entire wheat kernel) versus the white variety. Unless the outlet is focused on healthy eating, the probability is that white will overwhelmingly dominate whole wheat products.

For those concerned about the amount of energy, chemicals, and fertilizer that goes into wheat production, the consumption of the whole grain product essentially provides 50 percent more calories and more than double the nutrition per unit of such inputs. Also, the bran and other brown elements of the grain of wheat that is routinely fed to cattle in large amounts has arguably the most concentrated bundle of nutrition for the majority of humans of any natural grain product. It seems like such a rational choice to avoid the extra effort to produce white flour and extend that bushel of wheat by 50 percent.

I would like readers to study the enlarged photo of grains of wheat. To me it is a thing of beauty, and to think the healthy outer elements of the seed that you see in this picture are discarded for animal feed instead of human consumption seems even more of a folly. The next time consumers are faced with the choice between whole wheat products and their white counterparts, I suggest that they recall this image and, if nothing else think of the loss of such an attractive food.

Grains of Wheat

For rice, which along with wheat is the other global grain of choice, the same situation applies. White rice is overwhelmingly preferred to the more nutritionally robust brown products. The following table, also prepared by George Yu, tells a similar story to wheat.

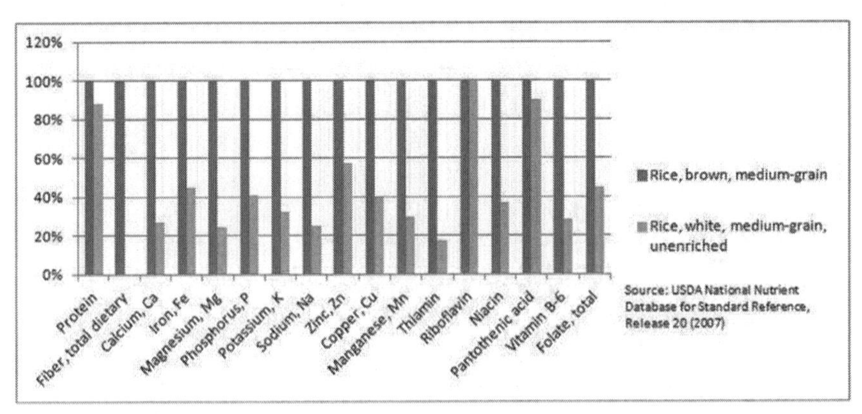

Nutritional Comparison of Brown and white Rice
Permission granted by George Y

As the situation with rice is a very close variation of the same theme discussed in wheat, there is little point in repeating the same issues of food abuse associated with substantial nutritional loss. It seems such folly that, in a world supposedly interested in healthy eating, the visual impact of the food is the driving force behind choice, more than any fundamental issue of how the grain might taste in its various forms.

The matter does not end here; other grains, such as barley. For consumers, pot barley, with only the inedible outer hull removed, is the more nutritious version. It is also known as barley groats or Scotch barley.

I cannot but wonder if anthropologists discovered an ancient society that discarded the most nutritious one third of their staple foods such as we do. It makes me wonder why our society cannot adopt a more enlightened path on this issue of über processed basic foods. This is only a consumer-driven folly and has little to do with the bogeyman of corporate greed because, as stated previously, bakeries and food processors stand to make just as much money off a brown product as they presently do from the white version.

Another food abuse is peeling certain fruits and vegetables such as apples and potatoes. Although I could not find as precise documentation comparing the peeled and unpeeled product as for wheat and rice, there is strong support in health-conscious circles for retaining the peels as a source of nutrients and fiber. Aside from nutrients, the first loss of peeling is the 10 percent or so of the entire product, because not only the skin is removed, but edible parts of the fruit or vegetable are cut away and discarded.

As far as the nutritional value is concerned, the following provides a detailed breakdown of potato peels.

One hundred grams (close to a quarter pound) of potato peel or skin has:[139]

58 kcal,
2.57 gm of protein,
12.44gms of carbohydrate,
0.1gms of fat,
2.5gm of total dietary fibre,
0.021mgs of thiamine,
0.038mgs of riboflavin,
1.033mgs of niacin,
11.4mg of vitamin C,
30mg of calcium,
3.24mg of iron,
413mg of potassium,
10mg of sodium.

It is an absolute abuse to discard such a food that has so much concentrated nutrition. An imaginative cook should be able to find some use for them. In our household peels even go into mashed potatoes, and once children are accustomed to it, the peels go unnoticed by not only our teenager and young adult children but also their friends. Then there are baked potatoes, home fries, and potato salad, all of which lend themselves to serving with the peel attached.

Another abuse is cooking, especially at high temperatures or for prolonged periods. In the overall pecking order of food preparation, raw is generally better than cooked. If cooking is desired, steaming is better than boiling, which in turn is less abusive nutritionally than frying or deep frying. Though most root vegetables and some others are hardly edible and need to be cooked before serving, keeping the peel intact prevents nutrients from leaching into the water if boiled.

Microwaving as a means to cook food is somewhat controversial, with many against the practice but mostly on anecdotal evidence. Aside from some unconfirmed concerns about the actual microwaves (e.g., the device being harmful to the operator), it would appear from 2003 research conducted by the National Center for Home Preservation, with support from the USDA and the Alabama Department of Food and Animal Sciences, that microwaving foods is more effective at maintaining nutrition than boiling. The following is from the abstract of the study.[140]

> *"The result showed that, compared to control samples, BWB [Boiled Water Blanching] lost 16% ascorbic acid, and 100% folic acid, thiamin and riboflavin while MWB (Microwave Blanching) lost 28.8% ascorbic acid, 25.7% folic acid 16.9% thiamin and 7.2% riboflavin."*
>
> *The results indicate that MWB is more effective in the retaining the selected water-soluble vitamins with the exception of ascorbic acid. This is also in congruence with earlier findings indication that microwave blanching is more effective in retaining nutrients in vegetables."*

Given that I could find no science-based research on the topic that provided contrary results, I suggest to readers that if they do not want to serve their vegetables raw, microwaving is a next best option.

Incidentally those who are concerned about the carbon footprint in the production and transportation of food as it makes its way to the fork should be aware that by some estimates, home food preparation (storage and cooking) requires 25 percent[141] of the entire energy output of the journey of food from the farm to the fork.

Looking again to grains, there is a relatively unimportant but interesting practice consumers have regarding corn on the cob that amuses me, regarding an unwitting abuse of food. A far as I am aware, immature corn is the only grain that we regularly consume in an unripened state and we have adopted it as a delicious and reasonably nutritious vegetable. Unbeknown to some, it has a fresh shelf life of only hours after the husk is removed. Yet in many supermarkets a bin is provided as a convenience to shoppers who want to avoid husking the corn in the kitchen. Unless the corn is cooked immediately when it gets home, nutritional and flavor loss starts to set in, as explained in the next paragraph.

Actually, "ripe" sweet corn is in a very delicate state of development just waiting to be fertilized before converting the sugar into the starch of a mature kernel. While the immature cob remains in the husk, the silk (strands that extend beyond the ear of corn) are individually attached to each kernel and serve as the route for the pollen to travel for fertilization. Once the husk and silk are removed, the kernel on the cob has no more chance for fertilization and, when combined with exposure to air, starts a process of accelerated deterioration compared to the situation if no husking occurs. The individual kernels will lose their plumpness and become somewhat shrunken more quickly than if they were unexposed. It is probable there will be nutritional loss, and definitely flavor will suffer. To demonstrate this point to fellow shoppers, think about purchasing a bunch of bananas, peeling them, and walking up to the checkout with a bag of rapidly deteriorating fruit.

Another abuse arises because most kitchens are generously stocked with sugars, salt, and oils that, if used in small quantities, provide a welcome enhancement to many foods. Unfortunately, as is all too often with deep frying or with excessive applications of sugar, empty calories are added to diets. Again cooks in the home kitchen start out with a healthy ingredient and thoughtlessly abuse it through the overuse of these three ubiquitous ingredients.

Much is written on the abuse that occurs before food arrives in the kitchen. This may be the case in too many instances, but so little attention is paid to the improvements that homemakers can achieve in the wise purchase and preparation of food. It seems quite contrary to logic that much is made on what processers can do to provide healthy food, but when control is in the hands of consumers, so little interest is shown by most to actually do something about it. Farmers, including local and organic growers, produce a wholesome product, and the distribution and processing system, in spite of some questionable practices, still offers a very wide range and huge quantities of wholesome and fresh products. For those who care about food, perhaps more attention be given to the home and what goes into the shopping cart environment, where they have full nutritional control. The energies that are extended on the quixotic rant by some against how food evolves prior to the point of purchase might be better expended where controls are absolute.

In conclusion, along with the hundred-mile diet, eating locally, and other such food consumption initiatives, it would be great if an "eat smart" movement would emerge that encouraged the best possible food purchase strategies and most advantageous preparation practices in the kitchen. There are many potential positives that such a concept might deliver—for example, less waste and more food available overall, greater nutrition, reduced carbon footprint to produce and deliver the required amount of food, and perhaps even less work for the preparation prior to the food appearing on the fork.

28.

Food Myths Busted

The media and other sources seem to have developed a herd instinct, and most claim that the journey of food from production on the farm to approaching the fork is, in many respects, broken. Though they clearly believe what they are saying or writing about, referencing each other, seldom do they go back to the fundamentals of official statistics or other sources such as peer-reviewed research, which often portrays a different reality. Some of the following are a repeat from elsewhere in this book, but perhaps readers will appreciate a condensed approach to some surprising facts.

Myth #1: The family farm is a thing of the past.

According to the 1997 US census, there were 1.9 million farms, of which 86.9 percent were designated "Individuals/family, sole proprietorship." Ten years later in the 2007 census, there remained a near identical 1.9 million and 86.5 percent.[142]

Myth #2: Corporate farms are taking over.

In 1997 there were 8,811 farms designated as "Non-family corporations (farms)," or 0.4 per cent of all American farms. This increased slightly to 0.5 per cent in 2007 at 10,237 farms. In other words, one farm out of two hundred is a corporate farm, and these would mostly be specialized dairy, beef, and poultry operations.[143]

Myth#3: Farms are getting larger and larger.

In 1997 49.2 percent of all US farms were 99 acres or less. In 2007 the percentage rose to 54.4 percent. During these same ten years, the average farm size decreased slightly from 431 acres to 418 acres.[144]

Myth #4: The farm population is declining.

Again comparing the two census years, there were 2.21 million farmers in 1997 and a nearly identical 2.20 million in 2007. For some it may come as a surprise that women as principle US farm operators increased by about 50 percent over the ten years, from 209,000 to 306,000. This means that one farm out of six is operated by a woman.[145]

To present a balanced point of view, this trend is the opposite in Canada.[146] I offer three explanations for this. With one-tenth the population but with farming spread out over even a wider distance from west to east than the United States, a much greater proportion of Canadian farms are distant from large population centers, where part-time employment is possible (which, as explained elsewhere in the book, is an economic necessity with smaller farming operations). Also related with closeness to urban centers is the opportunity for labor-intensive but profitable market gardens to supply fresh vegetables and fruit to nearby customers interested in eating local. Another consideration is the shorter Canadian growing season, which offers market gardeners a smaller window of opportunity to meet local urban demands for their produce.

Myth #5: Fertilizer use is increasing.

According to the USDA, corn growers' use 50 percent less nitrogen to produce a bushel of corn in 2005 compared to 1980. From 1990 to 2005, the reduction was still an impressive 17 percent less nitrogen fertilizer per bushel harvested. Over the same fifteen-year period, the reduced usage of phosphorous and potassium per bushel of corn produced was 28 and 20 percent, respectively.[147]

Myth #6: Organic farming makes a significant difference in the overall supply of food.

According to the 2007 census, there are 8,694 dedicated organic farmers with no conventional food production (less than 0.5 percent of all farms) and another 6,856 farmers who have certified organic production on part of their land. All organic farming utilizes 0.6 per cent of total agricultural land in the United States. In most industries this would be considered a mature and stable niche activity. Full recognition and support for organic farming dates back to the 1990 Farm Bill, and considering organic farming has been practiced for a lot longer than that, it is difficult to expect that the situation will change dramatically in coming years.[148]

Myth #7: Organic foods are produced without chemicals.

In the chapter on organic foods, I list the substances including "natural chemicals" that are permitted within the certification arrangement. For a practical insight on the reality of chemical use in organic food production, the following is quoted verbatim from page five of "The Agricultural Chemical Usage 2007 Field Crops" Summary Agricultural Statistics Board of the National Agricultural Statistics Board (NASS)of the USDA".[149]

> *"Organic Apples: Six Program States were included in the 2007 survey: California, Michigan, New York, Oregon, Pennsylvania, and Washington. However, due to data confidentiality rules the chemical application data for 5 of the 6 States surveyed were combined into one category called Other States (OS). Washington State was the only state with publishable data. Nitrogen, Phosphate, potash, and sulfur applications were made on 53, 25, 26, and 12 percent of the acreage, respectively. The rate per application per crop year for nitrogen was 131 pounds per acre; followed by 26 pounds of potash per acre; followed by phosphate and sulfur both being applied at a rate of 20 pounds per acre, respectively.*
>
> *There were no reports of herbicides applied to organic apple in the 6 Program States. Insecticides were used on 79 percent of the surveyed acres. Cyd-X Granulo Virus was the most commonly applied insecticide, on 56 percent of the acres; followed by Bt subsp. kurstaki on 53 percent of the acres. The pounds per acre per crop year for Cyd-X Granulo and Bt. Subsp. kurstaki were not publishable.*
>
> *Fungicides were used on 75 percent of the acreage. The more commonly used fungicides were Calcium polysufide, Sulfur, and Bacillus pumilus covering 66, 41, and 28 percent of the acreage, respectively. The average rate per application per crop year for Calcium polysulfide and Sulfur were 35.424 and 17.879 pounds per acre, respectively. The pounds per acre per crop year for Bacillus pumilus was not publishable.*
>
> *Other Chemicals were applied to 51 percent of the acreage, with*

> *Mineral oil being applied to 21 percent of the acreage at an average rate per crop year of 40.574 pounds per acre. Cytokinins and Butenoic Acid Hydro. were the next two most commonly applied Other Chemicals, at 11 and10 percent, respectively. The rate per application data for the active ingredient Cytokinins was not publishable. Butenoic Acid Hydro was applied at an average rate per crop year of 0.052 pounds per acre."*

I leave it to the reader to draw their own conclusions.

Myth #8: Food miles matter.

Simple arithmetic makes things interesting. According to the US Federal Railroad Administration of the Department of Transport, railroads move a ton of freight for a thousand miles on a little over two gallons of diesel fuel.[150] Thus it requires two gallons of fuel to move a ton of oranges the thousand miles from Orlando to New York City. By comparison, a typical automobile will get twenty miles per gallon in average city and country driving combined. Given that the USDA has fresh orange consumption at twelve pounds per person per year,[151] the energy requirement to deliver enough oranges for a family of four for two years is the same as driving a car for a mile. Sea shipments of food have even better food miles; trucking, which has the flexibility of moving the product from the orchard to market, will deliver about a six months' supply of oranges for four people for the energy equivalent of driving a few blocks. Only food delivered by air is worthy of any concern, but because this shipment method generally only pertains to the up-market, perishable, and out-of-season foods such as asparagus and raspberries, this should not be an overriding issue of a practical nature as they are easily identified by the individual who has a concern about food miles.

Myth #9: The food system is broken.

According to the FAO, the global percentage of undernourished has fallen from 33 percent of all humanity in 1969 to 16 percent in 2010, all while the world's population has increased from about four billion to nearly seven billion in 2010.[152] The number of adequately fed people more than doubled from a little over 2.5 billion to nearly 5.5 billion.

The world produces enough food to feed everyone. World agriculture produces 17 percent more calories per person today than it did thirty years ago, despite a 70 percent population increase. This is enough to provide everyone in the world with at least 2,720 kilocalories (kcal) per person per day.[153] Globally, the farming sector seems to be doing a fine job in keeping up with major population increases. What is broken however, is the socio-economic situation of individuals in all societies facing food security issues in the midst of more than adequate food supplies.

The USDA tracks food expenditures and has an excellent table entitled "Food Expenditures by Families and Individuals as a Share of Disposable Personal Income." From 1929, when this was first tracked, the percentage of disposable income spent on food was generally just under 20 percent until the early 1950s, when modern commercial agriculture is generally considered to have become widespread. Since then the decline has been very steady and reached only 5.5 percent of total income spent on food in 2008.[154] To personalize this rather stark statistic, the typical reader would have between three and four times the grocery bill today if it was not for modern agriculture supported by other improved efficiencies of the industry as food moves from farm to fork.

Myth#10: Organic foods are nutritionally superior to their conventionally grown counterparts.

As included in the chapter on organic foods, the September 2009 edition of the *American Journal of Clinical Nutrition* included a study entitled "Nutritional Quality of Organic Foods: A Systematic Review,"[155] which was conducted by researchers at the UK-based London School of Hygiene and Tropical Medicine. Their research involved a systematic review of 162 papers published from 1958 to 2008 on the comparison of the nutritional aspects of organic versus Non-organic foods. Of these 55 were of a satisfactory quality to be included in the final review. While the consensus was that organic produce and livestock products contained less nitrogen and more phosphorus than their conventional counterparts, the vitamin, mineral, and other nutritional aspects of both did not differ. The overall conclusion was that from the nutritional point of view, organic foods were not superior to conventionally grown and raised products. This systematic review, given the source and scope of the research involved, is difficult to dismiss. I will add, however, that the presence or absence of chemical residues was not within the terms of reference of this project.

29.

Is Information on Food Reliable?

One of the reasons for writing this book was that many of the things I read about food anywhere, from production on the farm and its journey all the way to the fork, were quite in contrast with my own knowledge of the subject. Given that the majority of individuals are exposed to the same media as I am, I thought it useful to provide an evaluation of such sources in an attempt to better equip readers in their own future research on food issues.

The Internet and media can be quite informative and factual, but many articles and discussions on food are based more on sensationalism or have an agenda with a very one-sided point of view that can be misleading. I therefore recommend to readers that whenever they suspect that the writer or presenter is not being totally objective, use their common sense and intelligence to ensure that the information is backed by independent and reliable third parties and is not someone's own opinion.

In early 2011 the US Society of Toxicology surveyed 937 of their members on several issues, including how they rate the media regarding their reporting on issues in their field of toxicology.[156] Given that these are the professionals who determine the level of harm, if any, through the exposure by humans and animals to chemicals or other substances, their opinions on how the media presents information on their field of expertise is quite telling. The following is from this report.

Rate the Media Coverage—

> *Toxicologists almost unanimously believe the media does a poor job covering basic scientific concepts and explaining risk.*

- *90% say media coverage of risk lacks balance and diversity*
- *97% say the media doesn't distinguish good studies from bad studies*
- *96% say the media doesn't distinguish correlation from causation96% say the media doesn't explain that "the dose makes the poison."*
- *Almost three out of four toxicologists believe the news media pays too much attention to individual studies as opposed to the overall evidence (74%), and to individual scientists as opposed to the broader community (73%)*
- *Over two out of three toxicologists (68%) believe the news media pays too much attention to studies put out by environmental groups, compared to only 27% and 18%who see too much media attention to studies by government and private sector scientists, respectively.*

These highly trained professionals condemn the media for doing a poor job of informing the public. For example, the difference between cause and correlation is all too often glossed over by the media. Those with strong views on a subject are very well aware of this and can garner a lot of attention by encouraging the media to use these without clarity regarding reality. To use an absurd example, the correlation of cancer victims who wear clothing is 100 percent, and yet to declare there is any cause here is preposterous. However, if the correlation was a little more opaque—such as nearly 100 percent of all farmers with cancer had used agriculture chemicals—perhaps those with an agenda might attempt to promote this correlation as evidence that agriculture chemicals cause cancer. The toxicologists would probably point out that nearly 100 percent of all farmers use chemicals, rendering correlation meaningless as far as cause was concerned. Their approach would probably be to compare cancer rates among farmers with that of the general population. Even here, if the correlation was a higher incidence of cancer for farmers, cause would not be proven, as it would be pointed out that farmers have extensive exposure to the sun and are perhaps more predisposed to skin cancer than the overall population.

Also, the concept "the dose makes the poison" perhaps requires clarification through example. Recently a friend sent me a blog with a host of articles on the dangers of chemicals in our foods. These were mostly anecdotal and did not seem to be backed by conventional research. One piece in particular caught my attention and reads as follows:

> *"Over 19% of commercial lettuce from major grocery store chains contained the pesticides DDT and DDE. Although research suggests these levels can affect humans, the EPA does not currently require chemical companies to test their pesticides*

> *for detailed immune system effects or subtle neurological effects (i.e. memory, concentration, personality, learning etc.)."*

This sounds pretty serious, but here is my take. It was an oversight not to indicate where, how, and by whom the survey was conducted. The author refers to these levels, but no mention is made on the concentration of these chemicals—only the incidence of detection.

When I consulted the website of the US Agency for Toxic Substance and Disease Registry,[157] I found the following:

> *"Summary: DDT (dichlorodiphenyltrichloroethane) is a pesticide once widely used to control insects in agriculture and insects that carry diseases such as malaria. DDT is a white, crystalline solid with no odor or taste. Its use in the U.S. was banned in 1972 because of damage to wildlife, but is still used in some countries. DDE (dichlorodiphenyldichloroethylene) and DDD (dichlorodiphenyldichloroethane) are chemicals similar to DDT that contaminate commercial DDT preparations. DDE has no commercial use. DDD was also used to kill pests, but its use has also been banned. One form of DDD has been used medically to treat cancer of the adrenal gland."*

From the same website I also found that *"half the DDT in soil will break down in 2-15 years, depending on the type of soil."*

Assuming a half-life average of ten years, this means that by 2012 the amount of DDT (the poison) in the soil will be one-sixteenth the level it was in 1972. Thus this researcher might get readings for DDT, but at what levels? My point here is why make such claims and concern the unsuspecting public, who cannot be expected to delve behind the statement as I have over a chemical that was banned four decades ago and that is nearly gone from the environment. To me, this is a tad malicious, but their statement certainly makes more dramatic reading than my counter argument.

Let me provide a couple more such examples of questionable presentation of facts. My local daily newspaper carried an article on the glycemic index (GI) of certain foods, which is the measurement of the impact of carbohydrates on the blood sugar, with desirable foods having a lower GI. In the article it was pointed out that whole wheat and white pasta both have a low glycemic index, with the whole wheat variety being a bit higher—which, in the opinion of the researcher interviewed, was not really significant. I consider this a sound observation, covered in two sentences in a six-hundred-word article.

Unfortunately the headline reads, "'Take a pass on whole wheat pasta'"—Not all 'brown foods' are healthier, says expert." Then the byline states, "Don't bother forcing whole wheat pasta down your throat: white pasta is just as good for you." Readers may not take the time to carefully read and understand that the researcher, who was interviewed, was not focusing on the merits of "brown" foods but rather the general concept of the GI, which is but one of many aspects of the complex issue regarding healthy foods.

A good friend of my wife, who is aware of our strong belief in whole wheat and brown rice products, accepted the headline at face value, as I am certain most readers would have done unless they took the time to carefully analyze the quotes in the article from the GI expert. Our friend's understanding was that the benefits of "brown" products were seriously in question, until I took the time to explain that as far as the GI is concerned, brown wheat products are neither advantaged nor disadvantaged compared to the white variety, and that nutrition was not even addressed in the article. The second unfortunate aspect of this misinformation was that the balanced and useful coverage of the GI was essentially lost due to this overlay of pointless sensationalism.

For a second example I turn to the book *Food Inc.*[158], which is a very readable account of some of the less savory aspects of the commercial food industry. The book is extremely negative in its one-sided portrayal of the subject, and to me, some of the contributors seem to rely on questionable "facts". Without some concept of the overall situation, one could gain the impression that the entire food industry is in dire straits. Fortunately the reader is given a very strong clue that objectivity is not to be expected, but to count on the entertainment value instead. There is an endorsement from *Variety* that is prominently displayed on the back cover: *"Food Inc.* does for the supermarket what Jaws did for the beach." As any thinking person knows, the movie was only intended to shock, and it did so in an outstanding fashion, but in no way was it considered a serious and objective scientific study of sharks.

Serious peer-reviewed articles and papers on food are sometimes difficult to both locate and understand, but such material may be worth a look on topics that are of interest to readers. Also, when an article seems sensational and one-sided, I encourage that the bonafides of the writer or presenter determine if they are a recognized expert in the field. Anyone who has a wealth of experience or an educational background in a subject will make certain that this is well-known to their audience. I suggest that readers keep this in mind as they build on the important understanding of the journey of food from farm to fork. It is a subject that is interesting, complex, and ever changing.

There is an old expression that goes something like this: "Figures don't lie, but liars will figure." To illustrate my point, I will use the comparison between whole wheat products and white flour in my chapter on food abuse. In this comparison of nineteen dietary components of flour, one is sodium, which is four times higher in whole wheat than in white flour. For someone who wants to make a claim that whole wheat is a challenged food, he or she can use a simple statement along the lines of, "Whole wheat has four times the sodium of white flour." The person would be technically correct, but the message would be entirely (and probably purposely) misleading to the point that consumers may doubt the overall evidence that whole wheat is really nutritionally superior to white flour products. As it transpires, whole wheat flour has only two milligrams of sodium per one-quarter cup, which is less than 1 percent of the recommended daily intake (in other words, all pure wheat products have insignificant levels of sodium).

It is almost a blood sport to declare the food system broken or to find some serious fault in what is an incredibly complex industry. First there are over a couple million farmers in North America that are producing a host of products that are dependent on a great many inputs, practices, and emerging technologies. Then there is the food processing and distribution system, with many thousands of players before the consumer even becomes involved. To be certain there are some practices by individual players—and even those that are industry wide—that could be improved upon. Unfortunately so much information dwells on these exceptions rather than providing a more balanced overview of the situation.

Then there are the willfully ignorant, who start out with the premise that the system is truly a mess and attempt to prove this single point by not only dwelling on the negative exceptions but proceeding to make up "facts" (that can easily be refuted by those who are professionally involved).

I encourage readers to be a diligent in determining junk information on food. That is often the difference between being informed and entertained.

30.

Biofuels and Food

Although this topic is not directly part of the farm-to-fork concept, biofuels increasingly compete for the same land and resources that traditionally were dedicated to food production. Thus it is important to understand both the current situation as well as the future of this relationship between food and fuel. I spent several years as an executive in the biofuels industry, and I feel reasonably well qualified to write on this topic.

There is a widespread conviction that the use of massive quantities of corn for the production of ethanol, and to a lesser extent soy beans for biodiesel, substantially contributes to hunger throughout the world. Though intuitively this seems to make sense, food insecurity usually is the consequence of such factors as poverty, inadequate infrastructure, or a failed state, with such poor governance and even instances of food aid being deprived from those who oppose a repressive regime. In reality, there is enough food in the world to go around, but getting it to all those who need it is the challenge.

There are concerns from time to time, and certainly at the time this book is being written, that the global stocks of grain and other foodstuffs are in a precarious state, but never in the post Second World War period have there been overall shortfalls in the amount of food available required to feed the world. This means that the world does not currently need all the corn and other grains that are dedicated to biofuel production, and thus they might as well be used for this purpose.

The following table covers actual corn production measured in terms of a thousand metric tons, and it demonstrates a dramatic increase in production that matches the expansion in US ethanol output in recent years. For skeptics on the availability of corn beyond ethanol production, 2008 was the record year for exports according to the USDA, and 2010 was the second highest year. In other recent years, the exports have been higher than the long-term average, which covers the pre-ethanol era. Critics also indicate that corn is being produced at the expense of other grains, of which wheat and soybeans are by far the largest contenders. Here exports again tell a different story. Soybean movements abroad set an all-time record in 2010 while wheat exports also had a record export year in 2008. Furthermore both commodities performed well above their long-term average in the past half decade. My point here is that the US capacity to produce food for domestic and export purposes exceeds the requirements of the ethanol industry. No one is going hungry because of ethanol production—at least yet.

US Corn Production

Change from

Year	(1000MT)	Previous Year
2002	227767	-5.64 %
2003	256229	12.50 %
2004	299876	17.03 %
2005	282263	-5.87 %
2006	267503	-5.23 %
2007	331177	23.80 %
2008	307142	-7.26 %
2009	332549	8.27 %
2010	316165	-4.93 %

(Source USDA)

When these total US corn production numbers are compared in the following graph[159] with the corn utilized for ethanol in recent years, it is quite apparent that overall corn production has expanded to meet the incremental requirements of the ethanol industry and yet provide traditional amounts for both domestic food purposes and exports.

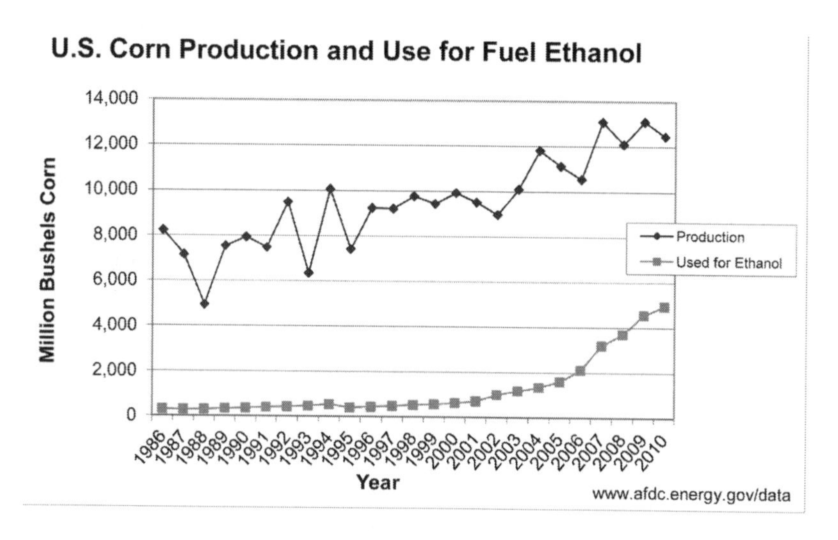

What is overlooked is the act that without the biofuels industry, corn and to a much lesser extent soybeans would not be produced in these quantities because there simply would be no market for the large amounts currently utilized by the biofuels industry. Thanks to this commercial activity, there is a reservoir of human food that would not exist if a large, ethanol-producing industry, particularly in the United States and Brazil (sugarcane is the feedstock of choice here) had not evolved over the past decade that can be drawn upon in the event of a serious disruption in global food production.

An analogy that might better explain this situation is a family that saves over the year for their annual summer holiday. Should the unforeseen happen and there is a substantial and necessary family expenditure, the holiday will have to be cancelled, but the integrity of the household remains intact. This is basically the same as with biofuels from food products. In other words, the amount of corn that is diverted to ethanol is "money in the bank" should global food supplies become insufficient.

In terms of human food versus fuel, what is the competitive dynamic? A bushel of corn weighs fifty-six pounds, whereas a pound of grain corn (not corn on the cob as a vegetable) contains about 1,700 calories, or about 85 percent of the daily requirements of a typical adult. Thus in terms of calories only, one bushel of corn is sufficient for approaching fifty days of basic food for a person. In round numbers, corn at $5.00 per bushel costs just ten cents per day to supply enough calories for a person to exist. Even at the peak price at $7.50 per bushel, the cost per day for the basic corn was only fifteen cents. I want to add that corn alone does not constitute a balanced diet, but it is a major staple food for millions. From this illustration I suggest that at these basic prices there is no reason for people to go hungry. To me there are other factors at play that cause food insecurity, but general availability and price are not critical factors.

Continuing with the arithmetic on food versus fuel, a 56-pound bushel of corn has a yield of 2.7 gallons of ethanol.[160] According to the Environmental Protection Agency of the United States, the average fleet mileage of all cars and light trucks in America is slightly over twenty miles per gallon of gasoline. Although ethanol has less energy content than gasoline, there are some studies that indicate that because of the higher oxygen content, there are some blends with gasoline that provide equal mileage. Thus I will give ethanol the benefit of the doubt and not discount the mileage; for the purposes here, a bushel of corn provides enough liquid fuel to move a car fifty-four miles. I should also add that there is a product called distillers grain that is left over from the ethanol production process, and it goes into cattle feed. However, as I gave ethanol the benefit on the mileage issue, and to keep matters from getting too complicated, I will not get into how much human food actually gets to the fork from this by-product of fermentation and distillation.

Without any moral judgment on my part, but rather an economic observation, it is interesting to note that the amount of corn required to propel a typical automobile for a mile would provide close to enough calories to feed a human for twenty-four hours. Here is where ethanol could face a serious economic problem when it comes to the comparable cost of crude oil for gasoline. A barrel of crude is forty-two gallons, of which about half is used to produce gasoline, and the balance is used for diesel and jet fuel plus other products with roughly the same value per gallon as gasoline. Thus the basic raw material costs, assuming a price of one hundred dollars per barrel of crude oil, is about $2.50 per gallon. A comparable raw material price for corn would be $6.75 per bushel. Corn ethanol also faces considerably higher costs of production and distribution than gasoline, which makes it particularly vulnerable to food-induced price increases for corn.

Thus should a global food shortage occur, I predict that the "food versus fuel" debate will become a moot argument. First, the economics of high corn prices in such a situation would probably squeeze the profit margins of ethanol producers to the point that it would be best to cease operations. Should the industry prevail because of government subsidies, then there would be the very real moral issue that does not exist currently. In this scenario, I cannot foresee that transportation's needs would trump the human hunger issue, particularly as gasoline would still be the dominant fuel in the marketplace. In practical terms this would mean that human hunger could be abated thanks to the corn supplies made available, compliments of the ethanol industry that inadvertently is doing much to make corn so abundantly available.

To elaborate on my comment on using gasoline, motorists can seamlessly and with no effort move to plan B for liquid fuels. The service station of choice will simply dispense pure gasoline, which in any event will still be the dominant part of the blend with ethanol in the future. As for the hungry of this world, there is no plan B such that fossil fuel provides for the car driver. Food from farms is the only fuel that works for these people.

The other major biofuel that may soon be a technical reality is cellulosic ethanol from biomass such as wood, grasses grown for this purpose, and agriculture residues such as corn stover and straw. In support of this transportation fuel, the US Energy Security and Independence Act of 2007 calls for sixteen billion gallons by 2022. Having spent the better part of ten years of my life scouring the United States, Canada, the United Kingdom, Germany, and France for biomass, I have serious doubts that the industry will ever achieve the level of production that legislators envisage.

Given that a ton of biomass will yield about eighty gallons of cellulosic ethanol, the sixteen-billion-gallon target requires some two hundred million tons of biomass. John Deere engineers estimate that this is equivalent to the tonnage of all food and fiber currently moved from farms in the United States.[161] Although the forestry industry has a long established history of delivering large quantities of biomass in an increasingly sustainable manner, the price of wood used for pulp and paper or construction purpose tends to be somewhat high for cellulosic ethanol producers. However, this source of biomass has no impact on agriculture or food production, and while it is possible to plant trees on land suitable for conventional farming, this is seldom the practice because forests can thrive in vast areas where food production is not possible. Unfortunately, in many forestry settings the economics are not particularly promising for cellulosic ethanol.

As far as agriculture is concerned, there will be some readily available material in surplus residues such as straw, and some opportunity for dedicated grasses on farmland unsuitable for the economic production of food crops. However, with anticipated increases in overall food prices, marginal land, which was so designated when prices were low, might easily become viable with improved economic conditions and advanced technologies. Also with rising fertilizer prices, agriculture residues are increasingly valuable when left on the soil to provide nutrients for next year's crop. This residue also provides protection from wind and water erosion as well as reduces the rate of evaporation, thus leaving the crops to thrive when rainfall is limited or irrigation water is challenged.

There is an opportunity for unproductive land to be utilized for the production of dedicated biomass crops, but this gives rise to issues of environmental impact and sustainability. For example, I once drove for a week in Florida, starting in Miami and then to Tallahassee, returning on a different route with the sole goal to scout the state for available biomass potential. There were promising pockets, but most of the state that was not already in agriculture appeared to be environmentally sensitive, and despite having the ideal climate for biomass production, the potential for the production of many millions of tons of this material could be a challenge. The situation is similar in most other states, and although land that is currently under the Conservation Reserve Program may pose some opportunities, here again there may be environmental issues in some areas; also, as this was all former farmland, an increased demand for food may divert the more productive elements of such holdings back into commercial food production.

The focus on the many cellulosic ethanol technology developers is on the know-how—there does not seem to be an appreciation that perhaps a couple hundred thousand farmers will need to be mobilized, each producing a thousand tons of biomass to meet the sixteen-billion-gallon target. Furthermore, this is not the only application for biomass, as local heat and power facilities are direct competitors for this feedstock.

Thousand Pound bales of Biomass (Straw)

On the other hand, there is at best only a modest appetite among most farmers to mobilize and produce the massive amounts of biomass that will be required. The US Congress, or at least the agricultural committees of the Senate and House, also appear to have a waning interest in promoting the widespread production of biomass by the farming community. Though there were several incentives supporting this activity in the 2007 Farm Bill, it appears that draft legislation being prepared for the 2012 version no longer will include such inducements for farmers and woodlot owners, for fiscal reasons.

In the above photo each bale weighs approximately half a ton and will provide sufficient material to produce close to forty gallons of cellulosic ethanol. Given that the available sustainable straw yield is typically one ton per acre, it requires slightly over 12 acres to produce a thousand gallons of this fuel[162]. To achieve even one billion gallons of the sixteen-billion-gallon target with straw biomass, twelve million acres would be required annually or one fifth of the entire US wheat acreage[163]. As the whet straw is already in demand for livestock, biomass for combined heat and power, and for sustainable agriculture practices my own experience is that the cellulosic ethanol industry can count on acquiring only a relatively small fraction of the total crop. For other biomass sources there are challenges that technology developers may not appreciate.

A bit of disclosure is probably in order here. Given that the call for sixteen billion gallons of cellulosic ethanol would require the efforts of the two hundred thousand farmers and perhaps woodlot owners, I established a consulting company in early 2010 to facilitate communities to become biorefinery-ready by taking the first steps to encourage their farmers to investigate the possibility of providing commercial quantities of biomass. I was invited to speak on the topic on a few occasions, but there seemed to be little interest on the part of community industrial development officials or farm groups to spend funds and actively mobilize for this purpose. My focus was on the United States, where I was well-known in many communities, and thus I had good expectations that there would be some opportunities for this initiative. I stay on top of the field and see little activity by others to actually marshal the critical mass of biomass providers that would be required to meet the future needs of an industry set to produce even a significant fraction of the sixteen billion gallons of cellulosic ethanol called for by 2022.

I am of the opinion that the corn ethanol industry is currently beneficial for global food security, inasmuch as it creates a significantly enhanced demand (and thus supply) for a food product during times of relative plenty, as been the case for the first decade of this century. Given that the calories required to feed an adult for a day comes close to the corn required to produce sufficient fuel to drive a typical car for one mile, the marketplace for human food will most likely outbid the one related to viable ethanol production when the occasion arises. The jury is still out regarding cellulosic ethanol, but even if it can mobilize sufficient interest among a great many farmers (which seems somewhat doubtful), it too will face future biomass supply issues wherever it competes on land that is also economically viable and suitable for food production.

31.

Feeding the Global Population

The future of food and the future of humanity are closely related, and this chapter is an attempt to describe the magnitude of this challenge. In this regard some interesting projections have been made on the impact that population growth will have on increased food demand. There are many variations on this theme, but Douglas Southgate has prepared one that is straightforward.[164] His projections range from a low population growth of about a billion in the next forty years to 7.87 billion, which would be accompanied by an average increase in annual consumption of 0.4 percent a year to account for improved conditions for today's malnourished—and of greater significance, enhanced consumption of animal proteins by the more prosperous. In this scenario of perhaps an unrealistically low population projection, the demand for food would increase by 59 percent. At the other end of the spectrum, he projects a population growth to 10.93 billion by 2050, or more than the estimate of 9.15 billion made by the United Nations Population Division[165] , but for this number he assumes less individual prosperity and therefore projects half the annual increase in per capita consumption, at 0.2 percent. In this case food production would need to increase by 99 percent—that is, we have to double the global farm output as it stands today.

Of special note is that the second scenario with the higher population growth is accompanied by half the increase in per capita food consumption and is at a level that would go little further than moving the billion or so individuals now in a malnourished state to one of being adequately fed. There is no provision for increased animal protein consumption, which is fast emerging as a major issue as the economies of India, China, and others prosper.

For any other industry, a worst-case growth scenario of 57 percent in the next forty years would be cause for celebration. For food producers the situation is positive, but for those who are concerned about global food security, this is a challenge of immense proportions. Not only is there the very real concern about hunger and severe human suffering and perhaps even more pervasive famine than the world now faces. This raises the prospect of population unrest and unstable governments, which would cause even more misery, is also a factor that could even reach the prosperous nations of today.

I do not necessarily see doom and gloom here, as there are several overall approaches that should be followed on an individual to a global scale that will make a difference. The most obvious is population control, but as this is such an emotional, social, spiritual, and moral issue, I leave this to others who are much better equipped to address this topic. However, it is very obvious that historically, as well as today, population increases go hand in hand with the availability of food, and an ever increasing population is the ultimate sustainability issue. For an excellent historic prospective on this subject, I recommend the book, *Empires of Food,* by Evan Fraser and Andrew Rimas.[166]

For the first time in history, the world has run out of substantial new lands. Though there is the occasional glowing report such as Brazil's Cerrado,[167] these are increasingly rare and probably do not make up for losses due to urban sprawl. Thus at best productive land is a fixed resource, so only improved yields are a possibility, and here technology has a lot to offer.

Perhaps the most promising development that could benefit the hungry are genetically modified crops. Critics may have some worthwhile points to make, but the uncompromising and often willfully ignorant claims by some that this technology is only harmful to humans and the environment are not helpful. Human lives are at stake. If there are deficiencies in this technology, the very science that brought them to the market should be called upon to find a remedy. Also, as with any human endeavor, there is an environmental downside when providing the necessities of life. Although much can be done to mitigate any negative impact, the reality remains that more food means there will be disruption to the environment. That is the unfortunate trade-off, but it should not be used as a reason to deprive the world of farming technologies that will feed the masses.

Here is where sensible approaches to sustainable agriculture have a very large role to play. Instead of a negative Luddite-type approach to advanced farming techniques, there needs to be a reasoned, science-based debate on what compromises will have to be made. The alternative to such compromise is the potential for widespread human misery and social unrest.

Yes, the world would be a much better place if one could put the genie back in the population bottle, but this will not happen. Intensive farming and the consequences thereof are a fact of life and will become even more of an issue in the future. Using science and technology to minimize negative outcome, and adopting the most sustainable alternative and yet meet global food requirements, will be the challenge.

One area where this approach will be particularly applicable will be aquiculture, or fish farming. The opponents to this practice are numerous and all to often uncompromising but do have some valid concerns. Again, the world would be better off environmentally if pen-raised fish were not required to feed humanity. As this is not the case, every effort should be made to develop the best technology that both mitigates environmental concerns and produces the maximum amount of healthy food. As mentioned earlier, new lands for farming are a thing of the past, but freshwater and saltwater remain a largely underutilized resource that could provide large quantities high-quality foods.

Outside of technology to produce more food, much can be accomplished by making better use of the food that we already produce. For starters, if some of the recommendations listed in the chapter on abuses of food were followed, the available food would go further. To repeat my favorite example, if all wheat consumed was the nutritionally superior whole wheat, this staple grain would provide nourishment for 50 percent more people. Given that wheat now provides 20 percent of the world's calories, this amounts to feeding nearly half a billion people.

Perhaps the largest loss of food that could be better utilized to meet population challenges is the production of animal protein. The ever-increasing per capita meat and dairy product consumption in both developed and emerging countries has an immense impact on the amount of food ultimately available for human consumption. I am not advocating anything approaching a ban on these food products, but there are pragmatic approaches that could make important differences. For starters, the wealthiest in the world are consuming more animal protein than is healthy for them. Here health professionals could do a better job in encouraging limiting the calories from meat, poultry, and dairy to appropriate levels. This promising situation, though it may not be easy to implement, should not be dismissed.

On a final note, given the immense challenge and opportunity facing the global food industry, the overwhelming concern of foodies for locally grown, organic, non-GMO, heritage tomatoes and other products seems misdirected. People should be concerned about the food they eat and the choices they have—I too share this awareness. What should not be overlooked is that the food production industry has a much larger task than catering to the prosperous elites.

32.

The Future of Food and Farming

There are several variables that will impact on the future of food and farming. The cumulative effect of these will be a challenge to predict, but I will discuss the dominant factors currently in play or on the horizon that will impact the global capability of humans to feed themselves. I will also provide my own views on this issue.

As was discussed in the previous chapter, there seems to be widespread consensus that the global population will expand from seven billion in late 2011[168] to nine billion by midcentury, and perhaps ten billion by 2100. Furthermore, with rapidly increasing prosperity in a host of countries, the demand for more animal proteins and processed food will increase with the result that more grains and other materials must be grown as inputs for these value-added products. These two trends mean that the finite farmland available today will be called upon to increase production by some 50–70 percent, which is substantially above the linear projection of the anticipated 30 percent increase in population. I place this very real issue ahead of energy requirements as the greatest challenge facing the global population.

This will place an incredible strain on the current agricultural sector, which already faces sustainability issues. To further acerbate the situation, the specter of climate change looms large. Though there is a lack of precision on the ultimate impact of ever increasing levels of carbon in the atmosphere, the most common scenario is for generally hotter and also dryer conditions. Moisture insecurity will be the result of higher temperatures, resulting in enhanced evaporation along with the possibility of reduced precipitation in some regions. For tropical areas this could prove to be very serious, while for more northern and southern climes the situation could be less of a problem and perhaps even provide conditions for enhanced food production, with longer growing seasons in regions that are challenged by short frost-free summers today. The main drivers to offset these conditions and a potentially devastating food security scenario, particularly on vulnerable populations, are technology and "best practice" enhancements for farmers, and improved governance.

Although inadequate to outright inept governance in several countries, and its negative impact on food production, is such an obvious impediment to feeding the world, the wherewithal to do anything about this complex issue is frustratingly limited. I am not optimistic that there will be a significant positive movement on improving governance, particularly in Sub-Sahara Africa, which holds so much promise to be a major producer and even exporter of food. Unfortunately governance issues have created a region with a population facing the most serious food security issues on the planet.

One development in this region that has promise to enhance food production significantly but carries huge negative social implications is the trend by the Chinese and others to purchase large tracts of land in governance-challenged countries and apply modern agricultural practices. For good reason this practice is frowned upon because it tends to disrupt existing rural societies and may do little to improve the diets of those who live close to such a farming structure. However, if the overall impact is to produce significantly enhanced quantities of desperately needed food, perhaps this practice should not be totally rejected. It is a long shot, but possibly international bodies with some clout could establish and enforce a code of best-practices conduct that would minimize the negative local social impact and perhaps provide some positive economic results. It will not be easy to establish such a system, but countries such as Zimbabwe, which has such food-producing potential and yet is unable to feed itself, would be better off with some form of enlightened application of capital and technology by offshore investors..

Of much greater promise is the practical role of enhanced technology and best practices to contribute substantially increased and sustainable food production and utilization.

I can think of no better example that no-till farming, which as previously discussed enables farmers to plant crops such as wheat and corn directly into the soil that remains undisturbed from the previous year's harvest. Not only is there a great savings in fuel and wear and tear on machinery, but water and wind erosion are greatly reduced, nutrient and chemical migration from the field is minimized, moisture is retained, and enhanced carbon sequestration occurs. The end result is increased yields, lower costs of production, and a substantial environmental dividend. Such tillage practices are already widespread in North America and elsewhere, and I expect that they will become universal.

A logical next step in this trend to minimize tillage is the development of perennial grains, particularly wheat that scientists are now developing. This crop is after all a grass, and if substantial harvests could be gleaned year after year from a single planting without any further disturbance of the soil, this would be major development in sustainable agriculture.

Another advanced technology that holds perhaps the most promise to feed the world is genetically modified crops. Though very much maligned, the benefits—particularly in the face of climate change—are quite overwhelming. As stated elsewhere in this book, genetic modification and other advanced plant-breeding technologies increase yields on a given piece of land without the need to utilize proportionally more inputs of fuel, fertilizer, and chemicals. Characteristics such as drought and salinity tolerance will be increasingly important as climate change unfolds, and thus it is essential that such food plants are developed in anticipation of their need so that the agriculture community can respond positively in a timely fashion to shifting climatic conditions as they occur.

Another application of genetic modification technology development that would bring immense sustainability benefits is the development of grains that fix nitrogen naturally in their root system, as do legumes (such as alfalfa or beans). Very simply, this would mean that the natural nitrogen from the atmosphere, instead of the hydrocarbon-derived version, would be available as fertilizer thanks to advanced root structures. Though this technology has challenges, scientists are confident that through advanced genetic applications, grains' reliance on synthetic nitrogen fertilizer will be substantially reduced.

My prediction on genetic engineering is that in a very few years, the general population will realize that a generation has gone by with billions of people eating a range of staple, genetically modified food products with few if any negative effects. Thus with this realization, along with a better understanding of the benefits, the negative attitude that so many currently have will fade as an issue of consequence.

Perhaps there will be residual rejection of this technology in such jurisdictions as Japan and the European Union, but it will be at the very real peril of increased food insecurity not only in less prosperous countries but also for their own citizens as they face food costs much higher than necessary. And then delete "if citizens and their governments do not fully accept GM foods and related ongoing technology developments." My concern that in the short run advances in this technology for crops in certain parts of the world such as Sub-Sahara Africa will be neglected because of the unfounded fear of GM technologies. Such a situation is a disincentive to plant scientists and provides little motivation to focus on the very real challenges of this region.

Aside from farming, consumers are becoming increasingly aware of best practices regarding how they purchase or prepare their food. The chapter "Abuses of Food" contains several concepts that could contribute to this trend.

I predict that there will be a food movement that develops and embraces the best possible practices regarding the sensible consumption of food. This will be an element of an increase in awareness by savvy populations in all aspects of food production, processing, marketing, and consumption

Another probable trend is that that global food prices will increase in coming years as populations grow and also become more prosperous. This will have a negative impact on food security, but it will enhance the prosperity of farmers and enable them to rise to the occasion to profitably embrace advanced and sustainable technologies. In turn, this will be an incentive to the food and plant research establishment to further push the frontiers of enhanced production technologies and practices.

I am reasonably optimistic that improved consumer awareness regarding maximizing the benefits of available food, the rapid adoption of new technologies, and perhaps improved governance in some jurisdictions will meet the dual challenges of an increasingly prosperous and rapidly growing global population, along with whatever climate change has in store for mankind.

It will also be a good time to be a farmer!

33.

Conclusions

Readers will have known well before they picked up this book that the topic of the journey of food from farm to fork is a complex one, with many diverse aspects on how it is produced, transported, processed, packaged, marketed, prepared, and consumed. I trust that the insights presented on these pages will enlighten some readers, but I also realize that some of my views (such as my general support for genetically modified foods, and my reservations about some aspects of the organic food movement) are in contrast with views held by many. Such differing views are healthy and form the basis of meaningful debate. In this regard I ask those readers who have zero tolerance to such innovations as genetically modified foods to modify their stance somewhat; while remaining against the practice in general terms, they can open their minds to study and understand what this technology can achieve in terms of feeding the world. In such debate one is encouraged to keep challenging the concept but also promoting best practices, as well as research and technology development, to make the food a better alternative than it may be today.

In a similar vein, I also invite readers to do their own research on those topics where they remain skeptical of the position that I have taken. I only ask that when engaging in such an exercise, readers do not purposely set out to prove me wrong by accepting only material that presents opposing views and by rejecting or ignoring supportive ideas. I have attempted to take a broad perspective on most issues and seldom come out either entirely positive or negative on any given subject. Given the complexity of the overall subject of food's journey from farm to fork, there are few unbending principles that preclude differing informed opinions. Thanks to this reality, I find the topic a dynamic one to study and write about.

Have an enjoyable and healthy relationship with your food!

Notes

1. There is no shortage of written and other material on negative aspects of the food industry. Thought there are some useful insights in this regard, many who contribute to this overall concept are, in my opinion, biased and prone to be less than thorough with the facts. I would invite readers who encounter an article, TV program, or book that is one-sided (perhaps to the extreme) to question the agenda of the writers or presenters and what credentials they gave as an expert in the flow of food from the farm to fork.

2. I possess a copy of the family tree of my Norwegian grandmother with the maiden name Rahm, which traces the family back to the 1400s. There are also documents on the Danish grandfather's family, Sorensen, going back a similar period of time.

3. These include Northwestern University, Evanston Illinois (scholarship for a semester in transportation economics), University of Alberta, Edmonton (BSc. in agriculture economics), and University of British Columbia, Vancouver (MSc. in agriculture economics)

4. I held the position of marketing director at Iogen Corporation from 1999 to 2009. The company specialized in the development and commercialization of technology to convert biomass into cellulosic ethanol as a transportation fuel.

5. http://www.ibm.com/smarterplanet/us/en/food_technology/ideas/

6. http://dictionary.reference.com/browse/farm+to+fork

7. For creationists I appreciate that this section has little relevance. Anthropologists are somewhat divided on the issue when "modern man" first evolved and generally place the period of 250,000–400,000 years ago. This lack of precision should not matter, as I am

only trying to illustrate how relatively recently in our overall history hunter-gatherers became primitive farmers.

8. As is the case on the origins of humans, precision on when people first started farming is subject to considerable conjecture by scholars who have several interesting theories. Ten thousand years seems to be a consensus, and thus I have adopted this to illustrate the fact that modern agriculture and its impact is an extremely recent phenomena in terms of total human history.

9. I use the term "common sense" here because I could find only speculation, but no peer-reviewed research, that the human body assumes famine during a crash diet and is inclined to create even more body fat whenever given the chance. However, there is ample evidence that only a small proportion of crash dieters can maintain a lower weight as a lifetime achievement, with most returning to at least their initial weight in the early months after the period of intensive dieting.

10. On this fact sheet provided below, the information on farm ownership can be located under the subheading: "Farm Characteristics." http://www.ers.usda.gov/statefacts/us.htm

11. The arithmetic goes like this: 50 tons is 100,000 pounds, or close to 1,700 bushels each weighing 60 pounds. Given that a bushel will bake into 90 loaves of whole wheat bread, this amounts to nearly 150,000 loaves, or sufficient calories (and a fair bit of useful nutrition) to feed 75,000 people for twenty four hours. Over an 8-hour perfect harvest day enough wheat is gleaned to feed 600,000 for the day or 1700 for a year. And then there are two combines.

12. See endnote 9. http://www.ers.usda.gov/statefacts/us.htm.

13. Considerable detail on Imperial is contained in their website: http://www.imperial-ne.com/

14. For Chase County: http://www.city-data.com/county/Chase_County-NE.html.

15. To acquaint readers with Birchhills: http://www.birchhills.ca/

16. I am providing these websites to give readers a sense of the vibrancy of these rural communities. For Jenni's Café: http://jennisnewgroundcafe.blogspot.com/

17. http://www.ers.usda.gov/statefacts/us.htm. On this fact sheet, the information on farm ownership can be located under the subheading "Farm Characteristics."

18. The following USDA web site from the 2007 and previous census years indicate the farming activity for new farmers as compared to the overall farming population. http://www.agcensus.usda.gov/Publications/2007/Online_Highlights/Fact_Sheets/new_farms.pdf

19. For an overview of the canola industry in Canada and insight into this increasingly important us crop, visit http://www.canola-council.org/

20. According to the US Bureau of Labor Statistics, there were 2,251,000 workers classified as "in the agriculture and related industries," while for "non agriculture industries" the number was 137,138,000. In other words, for every two workers producing food, there are about 130 workers elsewhere in the US economy. For further details, see http://www.bls.gov/news.release/empsit.t08.htm.

21. The number of loaves of bread per bushel is included as one of the "Quick Facts" on the website of the US National Association of Wheat Growers: http://www.wheatworld.org/

22. National Agricultural Statistics Service (NASS) USDA. http://www.nass.usda.gov/Statistics_by_State/Kansas/Publications/Crops/whthist.pdf

23. National Agricultural Statistics Service (NASS) USDA http://www.nass.usda.gov/Statistics_by_State/New_York/Publications/Statistical_Reports/01jan/fld10112.txt

24. The predominate wheat grown in Kansas is hard red winter wheat, which has a higher protein content than the higher yielding soft red winter wheat, the primary type grown east of the Mississippi. (This information came from a fact sheet that was prepared by the Texas Wheat Producers Board.)

25. The National Association of American Railways report that their members can move a ton of freight an average of 484 miles on a gallon of fuel. http://www.aar.org/Environment.aspx

26. From the company website www.Heinz.com

27. From the Almond Board of California website: www.almondboard.com.

28. National Agricultural Statistics Service (NASS) For Idaho production (http://www.nass.usda.gov/Statistics_by_State/Ag_Overview/AgOverview_ID.pdf) and for national output: (http://quickstats.nass.usda.gov/results/555669E1-0D78-3CBF-A9F3-525EF1414BCF?pivot=short_desc

29. McWilliams, James, "*Just Food*" Little, Brown and Company (2009) The author devotes the entire forest chapter to a thorough analysis on the relative insignificance of food miles.

30. For background on Rosie the Riveter and the impact that women made to the war effort in the 1940s, see: http://www.ohiohistorycentral.org/entry.php?rec=1676.

31. The Grocery Manufacturers 2007 publication "*INDUSTRY REPORT ON HEALTH ANDWELLNESS*" includes the following statement: "*92% of respondents have introduced or reformulated over 10,000 products and sizes offering many nutritional improvements*". http://www.gmaonline.org/downloads/research-and-reports/HealthWellness_07_FINAL.pdf

32. The Food Marketing Institute website is designed to inform consumers about this industry. http://www.fmi.org/facts_figs/?fuseaction=superfact.

33. According to the US Bureau of Labor Statistics, there were 2,251,000 workers classified as "in the agriculture and related industries"; http://www.bls.gov/news.release/empsit.t08.htm.

34. The US Food Safety and Inspection Service website contains a wealth of useful information for consumers, including food recalls and food safety suggestions: http://www.fsis.usda.gov/regulations_&_Policies/

35. This Canadian website is a variation of the same theme as the USDA counterpart, but for those wishing to develop an appreciation for the role that these two organizations have regarding food safety, a visit to both may be in order: http://www.inspection.gc.ca/english/toce.shtml.

36. The US Food and Drug Authority website is also worth a visit for concerned consumers: http://www.fda.gov/

37. Health Canada: http://www.hc-sc.gc.ca/index-eng.php.

38. "Cost-effectiveness of interventions to reduce dietary salt intake," Linda J Cobiac School of Population Health, The University of Queensland, Herston, QLD 4006, *Australia Heart* 96:1920–5.

39. James A Howenstine, MD, *A Physician's Guide to Natural Health Products That Work*, Pendhurst Books, January 2002

40. From the Diet Bites website: "There are 3,500 calories in a pound (of body fat). In order to lose OR gain one pound, an individual must either subtract or add 3,500 calories from their diet." http://www.dietbites.com/calories/calories-in-a-pound.html

41. An excellent overview of the issues regarding the taxation of sugary beverages appeared in the *New York Times:* "Proposed Tax on Sugary Beverages Debated," by William Neuman, published September 16, 2009.

42. These policies can be found as follows and are part of a much larger initiative on healthy living for the State of California. lihttp://www.publichealthadvocacy.org/_PDFs/beverage_policies/LocalPolicies_WaterSoda_Nov2010.pdf

43. Besides carryover, the following website includes statistics for several years including use for ethanol, livestock feed, human food, and exports. http://www.extension.iastate.edu/agdm/crops/outlook/cornbalancesheet.pdf

44. For details on issues relating to the farm safety net program, the US Congressional Research Service has provided an overview as an element of the 2012 Farm Bill: http://www.farmpolicy.com/wp-content/uploads/2010/09/CRSFarmSafetyNet2012FarmBill1-Sep10.aspx_.pdf.

45. For an official overview of the Common Agricultural Policy of the European Union, visit: http://ec.europa.eu/agriculture/publi/capexplained/cap_en.pdf.

46. USDA Economic Research Service: *Major Uses of Land in the United States, 2002* The following includes a complete breakdown of all land use. *"The United States has a total land area of nearly 2.3 billion acres. Major uses in 2002 were forest-use land, 651 million acres (28.8 percent); grassland pasture and range land, 587 million*

acres (25.9 percent); cropland, 442 million acres (19.5 percent); special uses (primarily parks and wildlife areas), 297 million acres (13.1 percent); miscellaneous other uses, 228 million acres (10.1 percent); and urban land, 60 million acres (2.6 percent). http://www.ers.usda.gov/Publications/EIB14/

47. The Conservation Reserve Program was established under the 1985 Food Security Act which is commonly known as the "Farm Bill."

48. Farm Service Agency Press Release No. 0671.10, "Conservation Reserve Program Celebrates 25 Years," Washington, Dec. 23, 2010.

49. For the United States: http://www.ducks.org. For Canada: http://www.ducks.ca.

50. Ibid.

51. "*An organism whose DNA has been altered for the purpose of improvement or correction of defects is considered to be genetically modified."* From "*genetically modified foods," Collins English Dictionary, Complete & Unabridged 10th Edition,* William Collins Sons & Co. Ltd., 2009.

52. Robin McKie, "*Genetic modification: Glow-in-the-dark lifesavers or mutant freaks*?" Guardian (UK) August 8, 2010.

53. Roger Highfield, "*'Spider-goats' start work on wonder web*," The Telegraph (UK), January 18, 2002.

54. Richard Gray, "Genetically Modified Cows Produce 'Human Milk,'" *Telegraph* (UK), April 13, 2011.

55. There is an excellent discussion as well on this site regarding GM foods and their production: http://www.ornl.gov/sci/techresources/Human_Genome/elsi/gmfood.shtml.

56. Stuart Wolpert, "*Dogs likely originated in the Middle East, new genetic data indicate*," UCLA Newsroom, March 17, 2010 http://newsroom.ucla.edu/portal/ucla/dogs-likely-originated-in-the-155101.aspx. This is a facts-based article that dog lovers should find interesting.

57. The Vincylopedia website provides an excellent short history of the devastation to the European vineyards by the phylloxera insect and

the recovery through grafting of the resistant American root stock: http://www.winepros.org/wine101/vincyc-phylloxera.htm.

58. The University of Minnesota Extension Service has a practical guide on how to graft fruit trees: http://www.extension.umn.edu/distribution/horticulture/dg0532.html.

59. Romans 11:17: "And if some of the branches be broken off, and thou, being a wild olive tree, wert graffed [old English for 'grafted'] in among them, and with them partakest of the root and fatness of the olive tree."

60. The FAO website has a somewhat technical but readable paper: "Molecular Markers and their Application in Cereals Breeding" by Viktor Korzun. http://www.fao.org/biotech/docs/korzun.pdf.

61. A background document on the various applications of molecular marker technology was prepared for an FAO conference in late 2003: "*Molecular marker assisted selection as a potential tool for genetic improvement of crops, forest trees, livestock and fish in developing countries,*" http://www.fao.org/biotech/C10doc.htm.

62. The February 2002 study had the title "*Genetically modified plants for food use and human health—an update.*" It can be found on the Royal Society website: www.royalsoc.ac.uk.

63. David Adam, "*GM research is needed urgently to avoid food crisis, says Royal Society,*" The Guardian (UK), October 21, 2009: http://www.guardian.co.uk/environment/2009/oct/21/gm-research-food. GM techniques will help crops survive harsher climates, as populations grow and global warming worsens, says the report.

64. The entire paper by the British Medical Association can be found at: http://www.bma.org.uk/images/GM_tcm41-20804.pdf.

65. Louise Gray, "*Chief scientist says GM and nanotechnology should be part of modern agriculture,*" The Telegraph (UK), January 6, 2010: http://www.telegraph.co.uk/earth/earthnews/6943231/Chief-scientist-says-GM-and-nanotechnology-should-be-part-of-modern-agriculture.html.

66. "*The Safety of Genetically Modified Foods Produced through Biotechnology,*" Toxicological Sciences Journal Volume 71, Issue 1, January 2003: http://toxsci.oxfordjournals.org/content/71/1.toc.

67. This site is not a one-on-one debate between two individuals on each side of the GM food debate. It is matching statements of one individual against peer-reviewed articles that were published as scientific documents in their own right and were not designed as a rebuttal to any particular statements made by opponents of genetic modification. http://academicsreview.org/reviewed-individuals/jeffrey-smith/

68. This quote on the nutritional benefits of GM rice came from the European Food Information Council article *"More iron and vitamin A from GM rice,"* published in November 1999. This organization looks upon GM technology as necessary and important to feed the global population. Their mandate is, "The European Food Information Council (EUFIC) is a non-profit organization which provides science-based information on food safety & quality and health & nutrition to the media, health and nutrition professionals and educators, in a way that promotes consumer understanding." http://www.eufic.org/article/en/food-technology/gmos/artid/iron-vitamin-a-gm-rice/

69. The mandate from the National Cancer Institute website: "*The National Cancer Institute (NCI) is part of the National Institutes of Health (NIH), which is one of 11 agencies that compose the Department of Health and Human Services (HHS). The NCI, established under the National Cancer Institute Act of 1937, is the Federal Government's principal agency for cancer research and training. The National Cancer Act of 1971 broadened the scope and responsibilities of the NCI and created the National Cancer Program. Over the years, legislative amendments have maintained the NCI authorities and responsibilities and added new information dissemination mandates as well as a requirement to assess the incorporation of state-of-the-art cancer treatments into clinical practice.*" http://www.cancer.gov/aboutnci.

70. From the website, their mandate reads, "*The National Association of Wheat Growers was founded more than 60 years ago by producers who wanted to work together for the common good of the industry. Today, NAWG works with its 21 affiliated state associations and many coalition partners on issues as diverse as federal farm policy, environmental regulation, the future commercialization of biotechnology in wheat and uniting the wheat industry around common goals.*" http://www.wheatworld.org/about-us/

71. The University of Dublin has done a thorough analysis of the decline in agricultural assistance by donor countries in the past few decades. http://www.tcd.ie/iiis/policycoherence/development-cooperation-trade-reform/trends-agricultural-aid.php

72. The National Agricultural Statistics Service (NASS) of USDA provides a treasure trove of historic and current information on all US crops and other farming activities. The specific information on corn and other grain yields can be found on the following website: http://usda.mannlib.cornell.edu/MannUsda/viewDocumentInfo.do?documentID=1047.

73. There is a near linear improvement in yield according to this USDA website: http://www.nass.usda.gov/Charts_and_Maps/graphics/cornyld.pdf

74. When associated with the National Association of Wheat Growers, I sat through several meetings addressing the issue of working with their Australian and Canadian competitors to jointly proceed with introducing GM wheat.

75. The National Organic Program of the USDA manages the certification and regulatory matters relating to organic foods that enter commerce in the United States: http://www.ams.usda.gov/AMSv1.0/nop.

76.

77. Ibid.

78. Economic Research Service, USDA: http://www.ers.usda.gov/Data/Organic/

79. Wency Leung, "*DNA Testing Will Bust Food Fraud and Contamination*," *Globe and Mail,* June 1, 2011.

80. Ibid.

81. Statistics Canada, 2006 Census of Agriculture: http://www.statcan.gc.ca/ca-ra2006/index-eng.htm.

82. Economic Research Service, USDA: http://www.ers.usda.gov/Data/Organic/

83. Gianessi et al., USDA, 2008 National Organic Production Survey: National Acreage and Crop Yields, March 2010, http://croplifefoundation.org/Organics/USDA%202008%20National%20Organic%20Production%20Survey%20Analysis.pdf.

84. The US Department of Energy graph depicting this relationship can be seen in the chapter on biofuels. It is also available at www.eere.energy.gov/afdc/data/index.html.

85. Hugh Martin, "Organic Crop Production," March 15, 2007, http://www.omafra.gov.on.ca/english/crops/field/news/croptalk/2007/ct-0307a2.htm.

86. The European Crop Protection Association website that includes this table: www.ecpa.eu/

87. The website "Extension," which is the product of the US Land Grant Universities, includes the following article on permissible chemicals for organic farmers: "*Approved Chemicals for Use in Organic Postharvest Systems,*" March 19, 2010, http://www.extension.org/pages/18355/approved-chemicals-for-use-in-organic-postharvest-systems.

88. For further information on the ethylene production, visit http://www.icis.com/v2/chemicals/9075778/ethylene/process.html.

89. University of Colorado "Some Pesticides Permitted in Organic Gardening" lists several pesticides, including Pyrethrins that organic growers are permitted to use. Coloradohttp://www.coopext.colostate.edu/4dmg/VegFruit/organic.htm

90. *Journal of Pesticide Reform*, Spring 2002, Vol. 22, No. 1. In addition to the quote from the EPA, this site has a list of the hazards of pyrethrins: http://www.pesticide.org/PyrethrinsPyrethrum.pdf.

91. Christine A. Bahlai, Yingen Xue, Cara M. McCreary, Arthur W. Schaafsma, Rebecca H. Hallett, "*Choosing Organic Pesticides over Synthetic Pesticides May Not Effectively Mitigate Environmental Risk in Soybeans,*" School of Environmental Sciences, University of Guelph, Guelph, Ontario, Canada. http://www.ncbi.nlm.nih.gov/pmc/articles/PMC2889831/

92. A. D. Dangour et al., *"Nutritional quality of organic foods: A systematic review,"* American Journal of Clinical Nutrition, Sept. 2009, http://www.ajcn.org/content/early/2009/07/29/ajcn.2009.28041.abstract.

93. The web site for the UK Foods Standards Agency is: http://www.food.gov.uk/

94. McWilliams, James. *Just Food: Where Locavores Get It Wrong and How We Can Eat Responsibly,* Little Brown and Company, 2009.

95. Ibid. Chapter 1, "Food Miles or Friendly Miles," pp 17–51.

96. This information on Canada's position in the world regarding population and food output can be found in the following website: http://www.statcan.gc.ca/pub/16-201-x/2009000/part-partie1-eng.htm.

97. The following website has a listing that includes all Canadian population centers' frost-free periods. Environment Canada is the referenced source. http://www.almanac.com/content/frost-chart-canada.

98. Colorado State University Extension fact sheet #7.220, http://www.ext.colostate.edu/pubs/garden/07220.html.

99. The website for the US Geological Service: http://pubs.usgs.gov/circ/2004/circ1268/

100. The Environmental Protection Agency provides a thorough overview of the Columbia River System on their website: http://yosemite.epa.gov/r10/ecocomm.nsf/Columbia/Columbia.

101. The British Columbia government has posted an overview of the history of the Columbia River Treaty: http://www.empr.gov.bc.ca/EAED/EPB/Documents/History%20ofColumbiaRiverNov139web).pdf.

102. USGS Study Documents, Water-level Changes in High Plains Aquifer, Released 2/9/2004, http://www.usgs.gov/newsroom/article.asp?ID=121.

103. The depletion issue is covered in some detail on the "Water Encyclopedia" website: http://www.waterencyclopedia.com/Oc-Po/Ogallala-Aquifer.html.

104. This website includes a calculation made by Jim Oltjen of the Department of Animal Science at U.C. Davis asserting that the requirements for a pound of beef to be 441 gallons of water. http://www.earthsave.org/environment/water.htm

105. The business activity of a sophisticated water district are provided in the Merced, California, website: http://www.mercedid.org/

106. Livestock and Poultry: *World Markets and Trade United States Department of Agriculture*, Foreign Agricultural Service , Circular Series DL&P 1-07, April 2007, http://usda.mannlib.cornell.edu/usda/fas/livestock-poultry-ma//2000s/2007/livestock-poultry-ma-04-11-2007.pdf.

107. The Science Daily web site has an interesting article that covers the history of the determination of the domestic chicken starting with Darwin. The conclusion is that the domesticated bird of today is genetically linked to both the red and gray junglefowl. http://www.sciencedaily.com/releases/2008/02/080229102059.htm

108. All USDA-approved meat and poultry labeling definitions can be found on the following website: http://www.fsis.usda.gov/factsheets/meat_&_poultry_labeling_terms/index.asp#3.

109. "*Our Big Pig Problem*," *Scientific American*, April 2011.

110. Renu Gandhi et al., "*Consumer Concerns about Hormones in Food," Program on Breast Cancer and Environmental Risk Factors"*, Cornell University Fact Sheet #37, June 2000.

111. the USDA website, which contains information on food labeling: http://www.fsis.usda.gov/factsheets/meat_&_poultry_labeling_terms/index.asp#15.

112. "World Population to Reach 10 Billion by 2100 if Fertility in all Countries Converges to Replacement Level," United Nations Press Release, May 2011.

113. Shannon Ryan et al., "*The History of the Northern Cod Fishery*," Department of Fisheries and Oceans (Canada), http://www.cdli.ca/cod/home1.htm.

114. *Scientific American*, February 2011, "*The Blue Food Revolution: Making Aquaculture a Sustainable Food Source.*" This article makes

a good argument that fish farming is not only sustainable and environmentally reasonably acceptable, but the only means to save depletion of wild fish stocks.

115. ibid

116. Sarah Spiker, "*Benefits of Precision Farming Technology: Reducing the Environmental Impact of the Farm*," January 27, 2009, http://www.suite101.com/content/benefits-of-precision-farming-technology-a92329#ixzz1LPs4QWky. This paper and the following one provide a useful review of advanced agricultural technologies.

117. Sarah Spiker, "B*asics of Precision Farming Technology: How Traditional Farming Is Being Replaced by New Technologies*," January 27, 2009, http://www.suite101.com/content/basics-of-precision-farming-technology-a92334#ixzz1LPv5WaSG.

118. Food and Nutrition Board (FNB), Institute of Medicine of the National Academies, "*Dietary Reference Intakes for Energy, Carbohydrate, Fiber, Fat, Fatty Acids, Cholesterol, Protein, and Amino Acids (Macronutrients),*" 2005, http://www.nap.edu/openbook.php?record_id=10490&page=1.

119. Council for Agricultural Science and Technology Task Force Report, "*Animal Agriculture and Global Food Supply*" July 1999, http://agrienvarchive.ca/bioenergy/download/anag.pdf.

120. Ibid., p. 1.

121. Ibid., p. 4.

122. The website "*All About Apples*" provides interesting details of the US apple industry: http://www.allaboutapples.com/facts.htm.

123. The Centers for Disease Control (CDC) have a death rate estimate that ranges from a low of 1,492 to a high of 4,938 with a median of 3037 for domestically acquired deaths (not from sources when travelling outside of the United States) from food-borne illnesses: http://www.cdc.gov/foodborneburden/2011-foodborne-estimates.html.

124. Other statistics on cause of death are available on the following CDC website: http://www.cdc.gov/nchs/fastats/acc-inj.htm.

125. The full paper is available at: http://www.tcd.ie/iiis/policycoherence/development-cooperation-trade-reform/trends-agricultural-aid.php.

126. *The Economist,* August 26, 2010, "*The Miracle of the Cerrado*". This article presents an excellent overview on how this previously unproductive area near the equator was transformed into a "breadbasket" with adapted temperate climate crops and advanced soil management technology. http://www.economist.com/node/16886442

127. Globe and Mail (Toronto) "*Mozambique Offers Farmland to Brazil*" *Report on Business, August 16, 2011*

128. Cotula Lorenzo, et al., "*Land Grab or Development Opportunity? Agricultural Investment and International Land Deals in Africa,*" 2009 FAO publication, http://www.new-ag.info/en/book/review.php?a=946.

129. From the CIA website: https://www.cia.gov/library/publications/the-world-factbook/geos/ch.html.

130. Ibid.

131. The percentage of total employment in agriculture as a share of all US jobs has declined from 3.5% in 1980 to 1.5% in 2007. http://www.tradingeconomics.com/united-states/employment-in-agriculture-percent-of-total-employment-wb-data.html.

132. USDA Food Security in the United States: Key Statistics and Graphics. The USDA projects that 85.3% of the American population is food secure, while 9.0% are classified as "low food secure" and a further 5.7% are deemed to be "very low food secure." http://www.ers.usda.gov/briefing/foodsecurity/stats_graphs.htm#food_secure.

133. Ibid.

134. A related USDA website contains global food insecurity information: http://www.ers.usda.gov/briefing/GlobalFoodSecurity/

135. Ibid.

136. The European Union has prepared an excellent overview of the history of the Common Agricultural Policy. http://ec.europa.eu/agriculture/publi/capexplained/cap_en.pdf.

137. The following paper provides a useful overview of the reasons for a farm safety net program plus an outline of a modified approach, which would modify some of the less desirable elements of the current program: Carl Zulauf, *"Designing a Safety Net for 21st Century Farming,"* Ag. Economist, Ohio State University, February 2011, OSU AED Economics (AEDE-RP-0134-11), http://aede.osu.edu/resources/docs/pdf/KB1OE3NQ-374E-Z3JV-YQHWAL2PVMVZ2B0Y.pdf.

138. Yu, George, *"The Simple and Natural Way to Vibrant Health" (2009) Xlibris*

139. http://www.bawarchi.com/health/queries34.

140. M. A. Osinboyne et al., *"Effects of microwave blanching vs. boiling water blanching on retention of selected water-soluble vitamins in turnip greens using HPLC,"* http://www.uga.edu/nchfp/papers/2003/03ifttur nipgreensposter.html.

141. McWilliams, James, *"Just Food"* Little, Brown and Company (2009) A discussion on the life cycle analysis of cooking food at home can be found on page 25.

142. The USDA website contains extensive census data on farm populations and farm types: http://www.ers.usda.gov/statefacts/us.htm.

143. Ibid.

144. Ibid.

145. Ibid.

146. Statistics Canada has comparable information on farm size and farm population, as does the USDA: http://www45.statcan.gc.ca/2008/cgco_2008_011-eng.htm.

147. During this period the national nitrogen application per acre was nearly constant (132 pounds per acre in 1990 vs. 138 pounds in 2005), while the US average yield increased from 118.5 bushels

per acre to 149.1 over the fifteen years. Yield data: http://usda.mannlib.cornell.edu/usda/nass/CropProdSu//2000s/2007/CropProdSu-01-12-2007.pdf. Fertilizer use statistics: http://www.ers.usda.gov/Data/FertilizerUse/

148. For the entire 2007 census: http://www.agcensus.usda.gov/Publications/2007/index.asp.

149. This survey covers the production of cotton, apples, and organic apples: http://usda.mannlib.cornell.edu/usda/nass/AgriChemUsFruits//2000s/2008/AgriChemUsFruits-05-21-2008.pdf.

150. Federal Railroad Administration Final Report, "*Comparative Evaluation of Rail and Truck Fuel Efficiency on Competitive Corridors*," November 19, 2009, http://www.fra.dot.gov/Downloads/Comparative_Evaluation_Rail_Truck_Fuel_Efficiency.pdf.

151. Susan L. Pollack, "*Characteristics of U.S. Orange Consumption,*" http://usda.mannlib.cornell.edu/usda/ers/FTS//2000s/2003/FTS-08-01-2003_Special_Report.pdf.

152. FAO World Hunger Education Service, "*2011 World Hunger and Poverty Facts and Statistics*," http://www.worldhunger.org/articles/Learn/world%20hunger%20facts%202002.htm.

153. Ibid.

154. USDA, "*Food CPI and Expenditures: Food Expenditure Tables," September 2008*, http://www.ers.usda.gov/briefing/cpifoodandexpenditures/data/Expenditures_tables/table7.htm.

155. A. D. Dangour et al., "*Nutritional quality of organic foods: a systematic review.*" American Journal of Clinical Nutrition, Sept. 2009, http://www.ajcn.org/content/early/2009/07/29/ajcn.2009.28041.abstract.

156. US Society of Toxicology, "*Are Chemicals Killing Us? Toxicologists Say Media Overstate Risks,*" http://www.toxicology.org/pr/AreChemicalsPR.pdf.

157. US Agency for Toxic Substance and Disease, Registry Toxic Substances Portal—DDT, DDE, DDD, http://www.google.ca/#hl=en&biw=1003&bih=567&q=US+Agency+for+Toxic+Substance+and+Disease+Registry+DDT&aq=f&aqi=&aql=f&oq=US+Agency+for+Toxic+Substance+and+Disease+Registry+DDT&fp=d5ad23689e277677.

158. Weber, Karl "*Food Inc*". Participant Media (2009)

159. This graph can be viewed at the Department of Energy website: www.eere.energy.gov/afdc/data/index.html.

160. Allen Baker, Steven Zahniser, "*Ethanol Reshapes the Corn Market,*" April 2006, http://www.ers.usda.gov/AmberWaves/April06/Features/Ethanol.htm.

161. On several occasions I presented at Farm Bureau workshops along with a representative from John Deere. Their calculation on the amount of biomass required to meet the sixteen-billion-gallon Renewable Fuel Standard was based on the simple calculation that it requires a ton of biomass to produce 80 gallons of ethanol, and thus 200 million tons for the sixteen-billion-gallon target. They had also totaled up the tonnage of all significant crops and arrived at a similar 200–million-ton figure.

162. Abengoa, one of the cellulosic ethanol technology developers predicts that the achievable yield will be 79 gallons a metric ton. http://www.greencarcongress.com/2011/02/abengoa-20110218.html

163. The USDA statistics indicate that the total acreage in wheat is typically close to 60 million acres. http://www.ers.usda.gov/Data/Wheat/YBtable01.asp

164. Douglas Southgate, *"Population Growth, Increases in Agricultural Production and Trends in Food Prices,"* Electronic Journal of Sustainable Development, http://www.ejsd.org/public/journal_article/13.

165. Foreign Affairs, "*The New Population Bomb*" – January/February 2011),

166. Fraser, Evan and Rimas, Andrew. "*Empires of Food*", Simon and Schuster, New York (2010)

167. *The Economist*, "*The Miracle of the Cerrado*," August 26, 2010, http://www.economist.com/node/16886442.

168. United Nations Press Release, May 3, 2011, *"World Population to reach 10 billion by 2100 if Fertility in all Countries Converges to Replacement Level,"* http://esa.un.org/unpd/wpp/Other-Information/Press_Release_WPP2010.pdf.

Index

D

E

F

G

H

I

J

K

L

M

R

S

T

U

V